Wheat-Flour Bread

Illustrations from 14th century manuscript
Medieval Health Handbook

IRON FORTIFICATION OF FOODS

FOOD SCIENCE AND TECHNOLOGY

A SERIES OF MONOGRAPHS

A complete list of the books in this series appears at the end of the volume.

IRON FORTIFICATION OF FOODS

Edited by

Fergus M. Clydesdale

Department of Food Science and Nutrition
University of Massachusetts
Amherst, Massachusetts

Kathryn L. Wiemer

General Mills, Inc.
Minneapolis, Minnesota

With the support of

1985

ACADEMIC PRESS, INC.
(Harcourt Brace Jovanovich, Publishers)
Orlando San Diego New York London
Toronto Montreal Sydney Tokyo

ACADEMIC PRESS, INC.
Orlando, Florida 32887

United Kingdom Edition published by
ACADEMIC PRESS INC. (LONDON) LTD.
24–28 Oval Road, London NW1 7DX

Library of Congress Cataloging in Publication Data

Main entry under title:

Iron fortification of foods.

(Food science and technology)
Includes bibliographies and index.
1. Food--Iron content. 2. Food additives. 3. Nutrition. 4. Iron deficiency anemia. I. Clydesdale, F. M. II. Wiemer, Kathryn L.
TX553.175176 1985 664'.06 84-24339
ISBN 0-12-177060-5 (alk. paper)

PRINTED IN THE UNITED STATES OF AMERICA

85 86 87 88 9 8 7 6 5 4 3 2 1

Contents

7. Breakfast Cereals and Dry Milled Corn Products

Ray Anderson

8. Iron Enrichment of Rice

John W. Hunnell, K. Yasumatsu, and S. Moritaka

B. Infant Products: Iron Fortification and Supplementation of Infant Formula and Products

9. Fortification of Infant Formula

Richard C. Theuer

10. Supplementation of Infant Products

George A. Purvis

Contributors

Numbers in parentheses indicate the pages on which the authors' contributions begin.

RAY ANDERSON (111), General Mills, Inc., James Ford Bell Technical Center, Minneapolis, Minnesota 55427

FRED BARRETT (75), Nutrition and Agribusiness Group, USDA, Office of International Cooperation and Development, Washington, D.C. 20250

JOHN L. BEARD (3), Nutrition Program, College of Human Development, Pennsylvania State University, University Park, Pennsylvania 16802

WERNER BEZWODA (55), Joint University/M.R.C. Iron and Red Cell Metabolism Unit, Department of Medicine, University of the Witwatersrand Medical School, Johannesburg 2193, South Africa

THOMAS H. BOTHWELL (55), Joint University/M.R.C. Iron and Red Cell Metabolism Unit, Department of Medicine, University of the Witwatersrand Medical School, Johannesburg 2193, South Africa

ROBERT CHARLTON (55), Joint University/M.R.C. Iron and Red Cell Metabolism Unit, Department of Medicine, University of the Witwatersrand Medical School, Johannesburg 2193, South Africa

G. COCCODRILLI, JR. (145), Nutrition and Health Sciences, General Foods Corporation, Technical Center, Tarrytown, New York 10591

CLEMENT A. FINCH (3), Division of Hematology, University of Washington, Providence Hospital, Seattle, Washington 98195

LARS GARBY (165), University of Odense, Institute of Physiology, DK-5230 Odense M, Denmark

LEIF HALLBERG (17), University of Göteborg, Department of Medicine II, S-143 45 Göteborg, Sweden

JOHN W. HUNNELL (121), Research and Development, Riviana Foods, Inc., Houston, Texas 77007

RICHARD F. HURRELL (39), Nestlé Products Technical Assistance Co., Ltd., Research Department, CH-1814 La Tour-de-Peilz, Switzerland

PATRICK MACPHAIL (55), Joint University/M.R.C. Iron and Red Cell Metabolism Unit, Department of Medicine, University of the Witwatersrand Medical School, Johannesburg 2193, South Africa

S. MORITAKA (121), Food Research Laboratories, Food Products Division, Takeda Chemical Industries, Ltd., Osaka, Japan

B. S. NARASINGA RAO (155), National Institute of Nutrition, Indian Council of Medical Research, Hyderabad-500007, India

JOHN PATRICK, JR. (31), SCM Metal Products, Chemical Division, SCM Corporation, Johnstown, Pennsylvania 15902

GEORGE A. PURVIS (139), Gerber Research Center, Gerber Products Company, Fremont, Michigan 49412

PETER RANUM (75), Pennwalt Corporation, Flour Service Department, Buffalo, New York 14240

N. SHAH* (145), Nutrition and Health Sciences, General Foods Corporation, Technical Center, Tarrytown, New York 10591

RICHARD C. THEUER (133), Research & Development, Beech-Nut Corporation, Fort Washington, Pennsylvania 19034

K. YASUMATSU (121), Food Products Division, Takeda Chemical Industries, Ltd., Tokyo 103, Japan

* Present address: Nutrition and Health Sciences, General Foods Corporation, Cranbury Technical Center, Cranbury, New Jersey 08512.

Foreword

Iron deficiency anemia is the most prevalent nutritional problem in the world today. It is estimated that at least one billion individuals are anemic because of insufficient iron. Most at risk are women of childbearing age, especially pregnant and lactating women, and young children. Iron deficiency anemia is found in all countries, but it is by far most prevalent and of greatest severity in developing countries. Moreover, the problem in these countries is further aggravated by the high prevalence of parasites that cause blood loss (e.g., hookworms and schistosomes).

Anemia in its mild form may affect an individual's ability to perform. This may manifest itself in reduced alertness and lethargy. As anemia becomes more severe, capacity to do work and performance are reduced. In pregnant women there is high risk of increased morbidity, premature delivery, and low birth weight. Studies suggest that iron deficiency may impair the body's immune system and the child's cognitive development.

Iron absorption in the intestinal tract is also reduced in vegetarian diets, which are often consumed in developing countries. Insufficient iron intake, low intestinal iron absorption, and increased iron losses thus combine to increase the prevalence of iron deficiency anemia in developing countries. Industrialized countries are not exempt from the deficiency; poor dietary practices in young children and limited food intake in young women, for aesthetic or other reasons, can produce iron deficiency anemia.

In the United States, Canada, the United Kingdom, Sweden, and several other developed nations, food fortification in general and iron fortification in particular have a history going back to World War II and earlier. However, in a number of other nations, particularly in Western Europe, fortification is not practiced at all because it is believed that fortification is a form of food adulteration. Nevertheless, there is a growing trend throughout the developed world for greater similarity in dietary patterns and practices, perhaps best illustrated by the rapid increase of "fast food" establishments, soft drink consumption, and institutional feeding. Furthermore, there is increasing international recognition that the nutritional quality of individual foods and of classes of foods is a fundamental food safety factor, whether in the context of developed or developing countries.

Given these current events, it seems reasonable to conclude that fortification with iron and other nutrients is likely to increase gradually in the developed world. This growth will be further stimulated as more nations make deliberate efforts to measure the nutritional status of their populations and to ascertain what residual deficiency syndromes persist. It is probable that they will find persistent iron deficiency in multiple age, sex, and physiologic groups, as is the case in the United States as determined by the periodic National Health and Nutrition Examination Surveys (NHANES). A further stimulus may well result from current research suggesting that significant decrements in both physical and cognitive performance may be associated with relatively mild iron deficiency in the absence of anemia. Also pertinent is the gradually developing consensus that, while iron supplementation (e.g., by use of tablets) may be suitable for situations limited in time and accompanied by regular medical supervision (e.g., during pregnancy), fortification of basic staples is the most efficacious, most innocuous, and cheapest mechanism for benefitting the diverse population segments at risk of iron deficiency. In addition, there is increased awareness of the fact that prudent use of iron fortification of common foods presents no danger of increasing the prevalence or severity of iron storage disorders such as hemochromatosis.

During the past decade, major progress has been made in identifying and interpreting the public health significance of iron deficiency, in understanding problems stemming from the wide variations in iron bioavailability as influenced by the specific chemical source of the iron and by the food vehicle used, and in the technology of adding iron to a variety of common foods. By no means are all the answers in, but progress is sufficient to warrant a detailed review as provided by this volume. It is important to keep in mind that, although application of fortification technology may vary from country to country, particularly between the developed and developing worlds, the principles developed from recent biomedical and technological research remain the same and apply to the entire world.

The U.S. Agency for International Development (AID) deserves particular recognition for its continuous support of the International Nutritional Anemia Consultative Group (INACG) which has played a key role together with the World Health Organization (WHO) in analyzing the major public health impact of iron deficiency world-wide, in developing methods of delivery and stimulating implementation of field trials, and in advancing the technological state of the art of iron fortification. Iron deficiency is the most prevalent deficiency disorder in today's world, affecting both developed and developing nations, and this timely volume should further contribute to alleviating the problem.

Edouard M. DeMaeyer
World Health Organization, Geneva
Allan L. Forbes
Food and Drug Administration, Washington, D.C.
Samuel G. Kahn
Agency for International Development, Washington, D.C.

Preface

With at least one billion human beings suffering from iron deficiency anemia, it is unnecessary to explain further the rationale that led to the development of this book. Recognizing that fortification of food with iron is the only feasible means to alleviate this problem, the International Nutritional Anemia Consultative Group (INACG) initiated this project.

The addition of iron to food can be traced to 4000 B.C., when Melampus, physician to Jason and the Argonauts, added iron filings to wine. In more modern times, iron fortification of food has been practiced for over three decades in many developed countries. However, the fickle bioavailability of iron has become well known as has its often deleterious effects on the food to which it is added. The major aim of this book is to discuss in detail the problems encountered with different iron sources in staple foods, beverages, condiments, and salt, as well as to provide a "how to" approach toward solving these problems in both developed and developing countries.

The first part of the book (Chapters 1 and 2) presents the reader with the basic problems of anemia—its prevalence, causes, and treatment—and discusses the effect of food on the availability of iron fortificants. The second section of the book is devoted to a comprehensive discussion of various iron sources, their interactions with food, and their bioavailability. This section is divided into three chapters, the first deals with commercially available elemental iron sources and the second with nonelemental iron sources used in food fortification. The third chapter in this section discusses experimental iron fortificants, their development, and use.

Section III pursues the critical area of product application. The first chapter in this section discusses wheat and blended cereal foods and contains the case history of a proposal for wheat fortification. The next chapter discusses breakfast cereals, most often associated with developed countries, and corn-based products. The extremely important area of iron fortification of infant foods is discussed in the next chapter, which is followed by a chapter on the fortification of various beverages including milk-based products, breakfast drinks, and soft drinks. The last two chapters review iron fortification of salt and condiments, respectively.

The book provides critically needed information for almost anyone, in any country, interested in fortifying food with iron and in treating iron deficiency anemia. It provides background for the social scientist or political reader, but is primarily intended for the technician involved with fortifying food. This work is unique in that it provides practical information after first discussing its theoretical basis. The examples given with the case histories provide enough information to allow a technician to begin a prototype iron fortification program almost anywhere in the world.

It is a tribute to the book and to the INACG that the contributors represent a worldwide who's who of iron researchers from academia, government, and industry. This has provided not only excellent science and technology but a multifaceted approach to the treatment of the subject as well. The contributors have all experienced the rewards and frustrations of working with this technology, and their experience is well utilized in this book.

We hope that the efforts of those involved will lead to a healthier world society. It has been a pleasure to have been a small part of this endeavor.

Fergus M. Clydesdale
Kathryn L. Wiemer

Acknowledgments

The publication of this book was made possible by a grant (DAN-0227-G-SS-1029-00) from the Office of Nutrition of the Agency for International Development to the Nutrition Foundation, which serves as the secretariat for the International Nutritional Anemia Consultative Group (INACG).

I

INTRODUCTION

1

Iron Deficiency

JOHN L. BEARD
Nutrition Program
Pennsylvania State University
University Park, Pennsylvania

CLEMENT A. FINCH
Division of Hematology
University of Washington
Providence Hospital
Seattle, Washington

I. DEFINITION

There are clinical and public health definitions of iron deficiency. The former is based on a deficit in essential body iron as recognized by the presence of anemia. The latter takes into consideration increased requirements for iron that may occur, for example, in pregnancy, and defines iron deficiency to include those individuals whose stores are inadequate to meet peak physiologic needs. An intermediate definition that has been widely used and that is practical from a diagnostic standpoint is to consider that iron deficiency is present when the supply of iron for body tissues becomes inadequate, regardless of whether anemia can be identified. These definitions differ only in the degree of iron depletion that they measure. This chapter applies to any of these definitions, although we are principally concerned with the prevention of iron deficiency rather than its treatment.

It is helpful to examine iron deficiency in relation to the various compartments of body iron (Bothwell *et al.*, 1979). Essential body iron is found predominantly (80%) in the form of hemoglobin. Myoglobin is the next largest fraction, followed by a variety of heme and nonheme tissue enzymes. Iron stores consist of

ISBN 0-12-177060-5

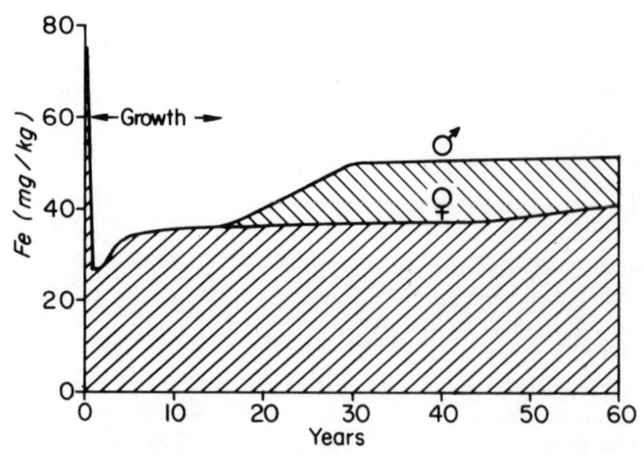

Figure 1. Body iron (mg/kg) content from birth through adulthood. The shaded area represents iron essential for normal physiologic proesses.

ferritin and hemosiderin, which are held in phagocytic cells and in hepatocytes to be made available when needed. The normal complement of total body iron depends on body size (Dallman *et al.*, 1980; Finch and Huebers, 1982). It amounts to some 300 mg in the neonate and gradually increases to 3–4 g in the adult (Fig. 1). The neonate is particularly rich in iron (100 mg/kg), and the surplus serves as a buffer against a limited iron intake during the first 3–4 months of life. Thereafter, body iron content drops to about 30 mg/kg and gradually increases to the adult level of 40 mg/kg for females and 50 mg/kg for males.

Iron balance involves the replacement of daily iron losses and whatever additional iron is needed for growth (Finch, 1976). For the adult male, iron requirements amount to 0.7–1.2 mg/day (about 14 μg/kg). Requirements of the menstruating female amount to between 1.2 and 2.5 mg/day (20–40 μg/kg), the greater range being due to the variability of menstrual blood loss (Hallberg *et al.*, 1966). Requirements for pregnancy, including the increase in maternal red cell mass, amount to an average of 4 mg/day, while net requirements (excluding maternal red cell mass) are about 2.5 mg/day (Bothwell *et al.*, 1979). Growth requirements in infancy are about 0.45 mg/day, and for the adolescent female and male during their adolescent spurt, are about 0.17 and 0.3 mg/day, respectively (Finch, 1976). From such figures, infancy and pregnancy may be identified as situations where iron balance is threatened, the former because of the very limited dietary iron uptake during infancy and the latter because of the high iron losses (Fig. 2).

The intestinal mucosa adjusts absorption according to body iron requirements (Bothwell *et al.*, 1958). If storage iron is increased, absorption correspondingly decreases. With depletion of iron stores, absorption is augmented. It has been hypothesized that this regulation depends on the relative contribution of various

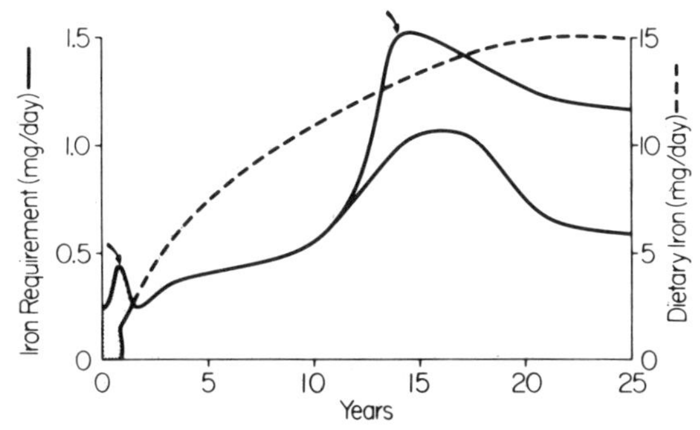

Figure 2. Iron balance as a function of dietary intake and body requirements. The solid black line represents requirements and divides at age 12 into females (upper line) and males (lower line). The dotted line represents a typical Western diet. Critical periods of actual and potential negative balance are denoted with arrows. [Taken from *Iron Metabolism in Man* by Bothwell *et al.* (1979).]

supply areas within the body versus the gut (Cavill *et al.*, 1975). When internal iron stores are reduced, a larger contribution is required from the gut to achieve iron balance. Thus, a relationship exists between iron stores and absorption as shown in Fig. 3. As iron requirements increase, stores are depleted and absorption increases until a new equilibrium point is reached. The relationship is illustrated by the lower storage iron and higher absorption found in women as compared to men, and this has been further characterized in blood donors (Finch *et al.*, 1977). One phlebotomy a year in men amounts to an additional requirement of 0.5 mg/day over a period of several years and will decrease iron stores by about 50%.

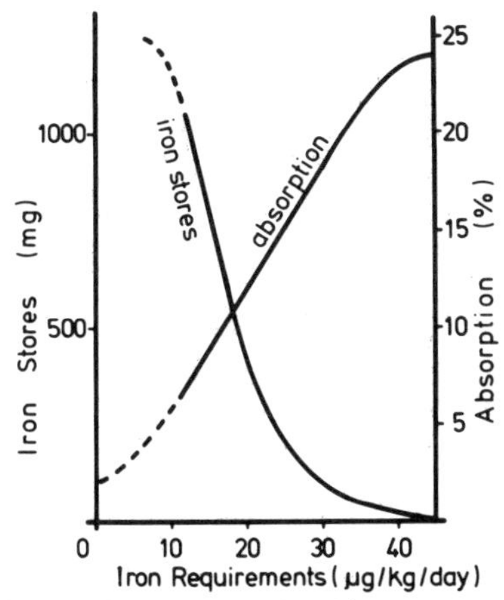

Figure 3. Relationship of absorption (percentage) and iron storage to daily iron requirements in humans. [Taken from Finch and Huebers (1982).]

The amount of dietary iron that may be absorbed under conditions of high requirement has been estimated in three different situations. One study involved severely anemic subjects treated for hookworm infestation in whom the gain in red cell mass was monitored (Roche and Layrisse, 1966). Another involved the phlebotomy requirements of patients with polycythemia vera (Finch *et al.*, 1950). The third was the calculated iron balance of high-frequency blood donors who were iron depleted but not anemic (Monsen *et al.*, 1984). These studies collectively suggest that there is a capacity to absorb 2–3 mg/day of iron from a normal diet in excess of physiologic losses other than menstruation. In this equation the nature of the diet becomes of considerable importance, and these figures apply only to individuals with favorable nutrition. Particularly in the economically deprived population, the composition of the diet may be such as to reduce greatly the amount of available iron (Hallberg, 1981). Indeed, in many parts of the world it must be assumed that available dietary iron is so limited as to produce iron deficiency in appreciable numbers of the female population whose iron requirements are purely physiologic (Hallberg, 1982). Other factors presumably enter into the complex equation of normal iron balance. For example, the extent to which absorption capabilities of different persons may vary is not known. It is also possible that additional, but unrecognized, iron losses may occur through occult gastrointestinal bleeding.

II. DETECTION

In the individual subjected to a continually negative iron balance, a predictable sequence of events will eventually occur (Bothwell *et al.*, 1979) (Fig. 4). Initially, iron stores will be mobilized to protect essential body iron, and absorption will increase. When stores are depleted, the iron supply to all body tissues as reflected in the plasma iron and transferrin saturation will become critically reduced, and there will be a progressive depletion of most essential body iron components, Ultimately, this depletion will be reflected in a decrease in circulating hemoglobin concentration.

The clinician usually enters the picture when anemia is detected. Investigations to determine its etiology frequently reveal a decrease in mean corpuscular volume as the first important clue. However, microcytosis is not specific for iron deficiency, and an additional laboratory evaluation is required involving the measurement of transferrin saturation, red cell protoporphyrin, serum ferritin, or marrow hemosiderin.

The same general approach has been used to establish the prevalence of iron deficiency among populations of normal individuals, that is, the initial identification of anemic individuals followed by the establishment of the etiology of the anemia. While this approach is quite appropriate for the clinician, it is far from

	Normal	Iron Depletion	Iron Deficient Erythropoiesis	Iron Deficiency Anemia
Iron Stores →				
Erythron Iron →				
RE Marrow Fe(0–6)	2–3+	0–1+	0	0
Transferrin IBC (μg/dl)	330 ± 30	360	390	410
Plasma Ferritin (μg/l)	100 ± 60	20	10	<10
Iron Absorption	normal	↑	↑	↑
Plasma Iron (μg/dl)	115 ± 50	115	<60	<40
Transferrin Saturation (%)	35 ± 15	30	<15	<10
Sideroblasts (%)	40–60	40–60	<10	<10
RBC Protoporphyrin (μg/dl RBC)	30	30	100	200
Erythrocytes	normal	normal	normal	microcytic and hypochromic

Figure 4. Sequence of changes in various parameters induced by gradual depletion of body iron content. [Taken from *Iron Metabolism in Man* by Bothwell *et al.* (1979).]

satisfactory in population studies. It is clear that *anemia is not suitable as an initial procedure for the recognition of iron deficiency.* Fully, one-half of individuals with iron deficiency have hemoglobin concentrations within the so-called normal range, and approximately half of the individuals with a hemoglobin concentration below the cutoff point for anemia have a "physiologically normal" hemoglobin concentration (Garby *et al.*, 1969; Cook *et al.*, 1971). So substantial an error is introduced by this primary separation based on anemia that prevalence figures for iron deficiency based on this approach have little meaning. If, on the other hand, measurements of iron status are first carried out to identify iron deficiency and then followed by a determination of serum hemoglobin to gauge its severity, it is possible to define the iron status of a population with some precision (Cook and Finch, 1979). Such an evaluation is shown in Fig. 5.

There are three valuable laboratory measurements by which iron status may be evaluated, and each of these has a somewhat different meaning (Cook, 1982). Probably the most useful measurement for populations with a high prevalence of iron deficiency, and the method for which the most extensive data are available, is the *transferrin saturation,* that is, the plasma iron divided by the total iron-binding capacity. The normal plasma iron fluctuates rather widely between 50 and 150 μg/dl while the total iron-binding capacity remains relatively stable at 330 ± 30 μg%. Normal transferrin saturation is 35 ± 15%. In the early stages of iron deficiency, the total iron-binding capacity increases in a fashion reciprocal to iron stores, but transferrin saturation remains unchanged. Only when stores are virtually exhausted do the plasma iron and transferrin saturation fall. When saturation drops below 16% there is an inadequate iron supply to support normal erythropoiesis (Bainton and Finch, 1965). At that point it may be presumed that

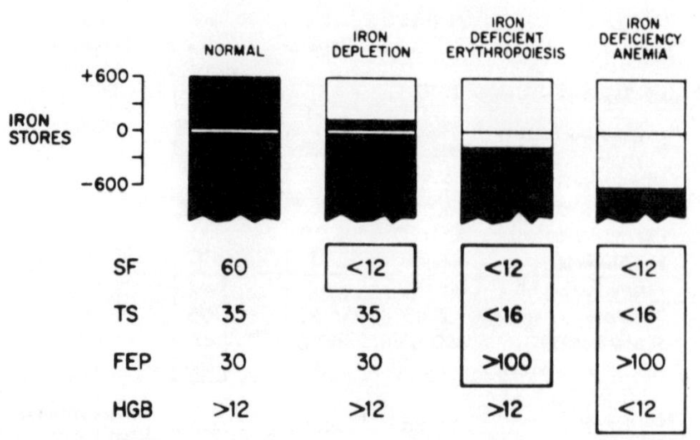

Figure 5. Parameters useful for assessing iron status of a population in a relation to body iron stores (mg). SF, Serum ferritin (μg/liter); TS, transferrin saturation (%); FEP, red cell protoporphyrin (μg/dl); HGB, hemoglobin (g/dl). [Taken from Cook and Finch (1979).]

depletion of essential body iron begins, but by the usual cutoff point for anemia only half of the patients with transferrin saturations of less than 16% will show anemia. A similar depression in transferrin saturation is found with the internal block in iron supply caused by inflammation. The total iron-binding capacity is of help in separating these two conditions, since it is often (but not always) elevated in iron deficiency but depressed with inflammation. Transferrin saturation can change rapidly, for example, within hours after the onset of inflammation. However, the increase of total iron-binding capacity is a protracted event reflecting the time required to mobilize iron stores.

Red cell protoporphyrin is another excellent means of detecting iron-deficient erythropoiesis (Labbe and Finch, 1980). Any decrease in iron supply to the individual red cell precursor results in an increase in the protoporphyrin precursor of heme. Several weeks of a deficient iron supply are required for a recognizable increase to occur in the circulating red cell protoporphyrin level. While this method is not as well standardized and its methodology must be carefully established in the individual laboratory, it provides a more stable index of iron-deficient erythropoiesis. It is also elevated in lead poisoning, although the increased absorption of lead is, at least part of the time, due to the associated iron deficiency. Protoporphyrin also measures something somewhat different from transferrin saturation, in that it takes into consideration the number of erythroid cells in the marrow (Langer *et al.*, 1972). For example, an individual with hemolytic anemia who has a plasma iron and transferrin saturation adequate for normal erythropoiesis may have a high red cell protoporphyrin, indicating a relative iron deficiency. This measurement, along with transferrin saturation, will be abnormal in inflammation, since in both conditions the internal block in iron metabolism critically decreases iron supply.

Serum ferritin provides a useful addition to the laboratory armamentarium for recognizing iron deficiency (Worwood, 1981). In general, the serum ferritin concentration reflects iron stores. The geometric mean in the adult male in developing countries is about 100 μg/liter, corresponding to iron stores of about 1000 mg. The corresponding value for adult menstruating females is 30 μg/liter, corresponding to iron stores of about 300 mg. For a more exact relationship between ferritin and the amount of storage iron, body size must also be considered. An impressive variation in serum ferritin exists between individuals, presumably reflecting variations in iron loss, in nutrition, and perhaps in absorptive capacity. When the ferritin concentration is below 12 μg/liter, an exhaustion of iron stores can be expected. In contrast to the two other measurements, serum ferritin is not decreased in inflammation but rather is disproportionately increased. Thus, the level of ferritin indicative of depleted iron stores in the presence of inflammation is increased to about 50 μg/liter. Despite this qualification, measurement of serum ferritin provides a convenient estimate of iron stores and is useful in detecting iron deficiency.

III. PREVALENCE

Prevalence of iron deficiency, as determined in any given population, will depend on the laboratory tests employed and what those laboratory tests define. For example, the general population might have a 25% prevalence of iron store depletion (ferritin concentration equal to or less than 20 μg/liter), a 12% prevalence of iron-deficient erythropoiesis (transferrin saturation less than 16%), elevated red cell protoporphyrin level (serum ferritin equal to or less than 12 μg/liter), and a 6% prevalence of iron deficiency anemia (iron-deficient erythropoiesis plus a depressed hemoglobin). Fortunately, earlier studies of prevalence of iron deficiency based on the initial detection of anemia are being replaced by studies that initially evaluate the adequacy of iron supply by one or more of the biochemical parameters previously discussed. This latter approach has identified an iron-deficient population without anemia (iron-deficient erythropoiesis or occult iron deficiency anemia) that roughly equals the number of individuals with iron deficiency anemia (World Health Organization, 1968). Several subsequent surveys have gone further by also evaluating iron stores employing the plasma ferritin determination, thus providing an even more complete profile of the iron balance of the total population (see Fig. 5). Other parameters, such as the mean corpuscular volume, provide useful information in cataloguing the type of anemia present, but this measurement will be abnormal in other disorders of hemoglobin synthesis, particularly thalassemia. Thus, specificity will vary with the population studied. Response to iron therapy would seem to be the final arbitrator of the presence of iron deficiency anemia, whether overt or occult. It was essential in validating the biochemical measurements of iron deficiency (Garby *et al.*, 1969). A valid therapeutic trial, however, is no simple matter and requires careful controls and a high degree of subject cooperation. Furthermore, there are other causes of variation in hemoglobin with time, which flaw even this approach to the etiology of mild anemia. For this reason, the therapeutic trial has been replaced by less demanding laboratory measurements.

Having selected a laboratory approach, standardized the methods, and evaluated their specificity and sensitivity in the setting in which they are to be used, one must then select a population that will meet the objectives of the study. A number of variables are known to exist that may need to be controlled; these include the economic status and racial origin of the people selected. Additional complications may relate to environmental factors that are different in the samples selected as compared to the general population. Major differences in prevalence also exist at different stages of life and between the two sexes. In a population whose menstruating females show a prevalence of 20% of iron deficiency with or without anemia, the equivalent for adult males may be only 3% (Cook *et al.*, 1976). These figures may be much higher in regions of the Third World as a result of a combination of poverty and increased blood loss through parasitic infections, but a differential between males and females will remain. Thus, in respect to both prevalence and severity, the selection of the adult menstruating female population provides a more sensitive index of iron deficiency. While the prevalence may be even greater in infants, the special dietary practices (breast-feeding, special fortified diets, etc.) make this a special problem. Pregnant women are more vulnerable than the adult menstruating female population, but the prevalence of iron deficiency is almost too high. The degree of anemia is also affected by fluid retention, and, particularly in developed countries, the pregnant female often takes iron supplements.

If the purpose of the survey is to examine the adequacy of the general diet with respect to iron balance, it seems best to examine the iron status of the adult female. In a number of studies carried out on populations of adult females throughout the world, the prevalence of iron deficiency varies from about 10 to well over 50%, the prevalence of iron deficiency anemia from 6 to perhaps 50%, and the prevalence of iron stores inadequacy from 35 to nearly 100% (Bothwell *et al.*, 1979; World Health Organization, 1968). In general, developing countries show the higher prevalence, the consequence of both hookworm infestation and a low availability of dietary iron. The number of individuals with iron deficiency is in the hundreds of millions.

IV. CAUSES OF IRON DEFICIENCY

From the preceding sections it is apparent that the human is able to withstand daily iron losses of only 2–3 mg. Amounts of greater magnitude, whether phys-

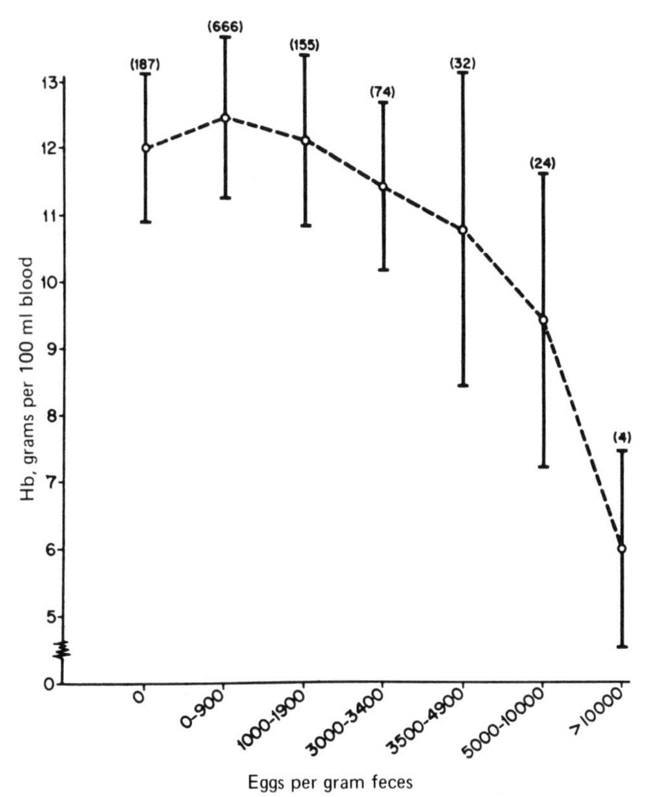

Figure 6. Effect of parasite load (number of hookworm eggs per gram feces) on hemoglobin concentration. The numbers of cases are given in parentheses, and vertical lines represent 1 SD. [Taken from Roche and Layrisse (1966).]

iologic or pathologic, will deplete iron stores and eventually lead to a decrease in essential body iron. Excessive menstrual blood loss in the female and repeated pregnancies result in iron deficiency. In addition, there are a multitude of causes of pathologic bleeding, usually from the gastrointestinal tract or uterus. Of particular importance in this respect is hookworm infestation, which is so common among shoeless segments of the world's population that infestation may be regarded as a way of life (Roche and Layrisse, 1966). Blood loss is proportional to the hookworm load (Fig. 6), and quantitative egg counts are essential in determining the importance of this source of blood loss. The physician must occasionally think of causes of iron loss other than bleeding and pregnancy, including such rare causes as hemosiderinuria in patients with intravascular hemolysis and pulmonary siderosis where the iron of extravasated red cells is virtually unavailable (Table I).

Reduced iron absorption can also lead to iron deficiency, but pathologic causes of malabsorption are rare and virtually limited to sprue sufferers and to

Table I
COMMON CAUSES
OF IRON IMBALANCE
IN HUMANS

Increased iron loss
Physiologic
Increased menstruation
Pregnancy
Blood donation
Pathologic
Bleeding
Hemosiderinuria
Decreased absorption
Inadequate dietary iron
Malabsorption

patients in whom gastric surgery has been performed. The major cause of reduced absorption is the nature of the diet itself. Diets containing no meat and little ascorbic acid have very limited iron availability and can lead to iron deficiency (Hallberg, 1981; Cook and Finch, 1975).

Since iron deficiency is usually the result of a poor diet combined with increased iron loss, it is not always clear what should be labeled *dietary* iron deficiency. For example, the menstruating woman who becomes anemic will usually lose more menstrual blood than most other women. Some estimate of an appropriate amount of available dietary iron must be set, and this amount of iron loss can be the dividing point between dietary iron deficiency and blood loss anemia. If we are satisfied with the status quo of the adequate diet of most developed countries, we have a diet that can provide about 3 mg/day of iron to the iron-depleted subject. Anyone who has iron deficiency with losses of less than that should be assumed to have dietary iron deficiency. Such a definition would almost certainly include most of the world's iron-deficient population. However, if we accept the present diet as adequate, we must also acknowledge that the iron balance that results is inadequate to meet the requirements of pregnancy in approximately one-third of adult women.

V. HEALTH SIGNIFICANCE OF IRON DEFICIENCY

There are difficulties in the translation of laboratory measurements of iron deficiency into their health equivalents. Certainly there is ample evidence from experimental animal studies to indicate the limits placed on work performance by anemia. Physiologic measurements in humans likewise show how oxygen transport and work performance are compromised by severe anemia (Dallman, 1982). For people engaged in strenuous physical activity for their livelihood, iron deficiency is translated into economic loss.

It has been assumed that anemia is the principal cause of disability in iron deficiency, and this is probably so. However, it is also recognized that there is a general depletion of virtually all iron compounds in iron deficiency (Dallman, 1974), and the possibility therefore exists that other effects due to critical enzyme depletion may occur. In the experimental animal rendered iron deficient but exchange transfused so as to eliminate anemia, depletion of mitochondrial enzymes occurs that, in itself, results in muscle dysfunction and impaired work performance (Finch *et al.*, 1979). While it does not seem likely from preliminary data that this is an important feature of iron deficiency in humans, the observation does suggest that such relationships need to be examined. Of possibly greater clinical relevance is the effect of iron deficiency on the maintenance of body temperature (Dillman *et al.*, 1979). Martinez-Torres *et al.* (1984) have shown that individuals with iron deficiency anemia require greater oxygen consumption to maintain body temperature during cold exposure. This could lead to negative caloric balance in individuals whose dietary intake is already restricted. Related studies in animals demonstrate alterations in thyroid hormone as well as catecholamine metabolism (Dillman *et al.*, 1980; Beard *et al.*, 1984). Behavioral changes have been reported to be associated with iron deficiency (Pollitt and Leibel, 1982). It will obviously be difficult to match the many biochemical alterations in iron deficiency with their possible functional consequences, but there is no longer need to doubt their presence. It seems reasonable, therefore, to regard iron deficiency as a generalized disorder whose presence has the potential, at least, of imposing limitations on a variety of bodily functions.

VI. METHODS OF PREVENTING IRON DEFICIENCY

In the ideal world, hookworm would be eradicated and the diet would contain sufficient meat and/or ascorbic acid to create a potential supply of 3–5 mg/day of absorbed iron. In the male without pathologic blood loss and with normal regulation of absorption, this would probably create stores of slightly over 1000 mg, and in the average menstruating female, stores of approximately 500 mg. Even the needs of pregnancy would usually be met by existing iron stores with subsequent gradual rebuilding of storage iron in the postpartum period. In such a setting, iron deficiency should be far less frequent and would be related to pathologic blood loss.

Unfortunately, there is no practical way of eradicating hookworm infection in the near future, nor is it likely that the general nature of the diet of the very population most affected will be modified so as to increase the availability of iron. The expense and impracticality of dealing with individuals on a case-by-case basis by the physician is likewise recognized. The solution to the problem is clearly in the domain of preventive medicine, and different strategies have been suggested (World Health Organization, 1975). In areas of high prevalence and severe deficiency, the administration of iron to certain target groups, such as young children and pregnant and menstruating women, has seemed desirable. Critical to the supplementation approach is the ability to reach the affected population. This implies an organized health service to implement the program, to educate the people as to its importance, and to maintain a continued surveillance of its effectiveness.

A nutritional approach whereby available dietary iron is held at an adequate level is the ultimate solution. Food iron fortification has been applied most effectively in infancy, through use of a specially formulated diet. There is good reason to believe it is effective (Hallberg, 1982). However, such fortified formulas are not available to the most needy infant population. A still greater problem is the improvement of dietary iron availability for the adult population. While fortification programs have been attempted in the past, there is little documentation of their effectiveness. More recent studies provide a much more substantial base from which to approach food fortification, and there is reason for optimism concerning its potential effectiveness. However, there are many practical problems. Fortification could be most easily accomplished in developed countries where there is a lesser need. Strategies for fortification will undoubtedly need to be different, depending on the special opportunity provided by the food pattern of the people. It is necessary to find a suitable vehicle—such as flour or salt—that is processed centrally but that will reach a large portion of the population. Alternatively, one may attempt to improve the availability of iron already present in the diet by adding some substance that will increase food iron availability, such as ascorbate. These seemingly logical solutions to the problem of a deficient supply of available iron are beset with various problems, which will be addressed in the latter part of this book.

REFERENCES

Bainton, D. F., and Finch, C. A. (1965). The diagnosis of iron deficiency anemia. *Am. J. Med.* **37,** 62–70.

Beard, J. L., Green, W., and Finch, C. A. (1984). Effects of iron deficiency anemia on thyroid hormone levels and thermoregulation during acute cold exposure. *Am. J. Physiol.* **247,** R114–R119.

Bothwell, T. H., Pirzio-Biroli, G., and Finch, C. A. (1958). Iron absorption. 1. Factors influencing absorption. *J. Lab. Clin. Med.* **51,** 24–36.

Bothwell, T. H., Charlton, R. W. Cook, J. D., and Finch, C. A. (1979). "Iron Metabolism in Man." Blackwell, Oxford.

Cavill, I., Worwood, M., and Jacobs, A. (1975). Internal regulation of iron absorption. *Nature (London)* **256,** 328–329.
Cook, J. D. (1982). Clinical evaluation of iron deficiency. *Semin. Hematol.* **19**(1), 6–18.
Cook, J. D., and Finch, C. A. (1975). Iron nutrition. *West. J. Med.* **122,** 474–481.
Cook, J. D., and Finch, C. A. (1979). Assessing iron status of a population. *Am. J. Clin. Nutr.* **32,** 2115–2119.
Cook, J. D., Alvarado, J., Gutnisky, A., Jamba, M., Labardini, J., Layrisse, M., Linares, A., Loria, A., Maspes, V., Restrepo, A., Reynafarje, C., Sanchez-Medal, L., Velez, H., and Viteri, F. (1971). Nutritional deficiency and anemia in Latin America: A collaborative study. *Blood* **38,** 591–603.
Cook, J. D., Finch, C. A., and Smith, N. J. (1976). Evaluation of the iron status of a population. *Blood* **48,** 449–455.
Dallman, P. R. (1974). Tissue effects of iron deficiency. *In* "Iron in Biochemistry and Medicine" (A. Jacobs, and M. Worwood, eds.), pp. 437–475. Academic Press, New York.
Dallman, P. R. (1982). Manifestations of iron deficiency. *Semin. Hematol.* **19**(1), 19–30.
Dallman, P. R., Siimes, M., and Stekel, A. (1980). Iron deficiency in infancy and childhood. *Am. J. Clin. Nutr.* **33,** 86–118.
Dillman, E., Johnson, D. G., Martin, J., Mackler, B., and Finch, C. A. (1979). Catecholamine elevation in iron deficiency. *Am. J. Physiol.* **237,** R297–R300.
Dillman, E., Gale, C., Green, W., Johnson, D. G., Mackler, B., and Finch, C. A. (1980). Hypothermia in iron deficiency due to altered triiodothyronine metabolism. *Am. J. Physiol.* **239,** R377–R381.
Finch, C. A. (1976). Iron metabolism. *In* "Nutrition Reviews, Present Knowledge in Nutrition" (D. M. Hegsted *et al.,* eds.), pp. 280–289. Nutrition Foundation, New York.
Finch, C. A., Cook, J. D., Labbe, R. F., and Culala, M. (1977). Effect of blood donation on iron stores as evaluated by serum ferritin. *Blood* **50,** 441–447.
Finch, C. A., Gollnick, P, D., Hlastala, M. P., Miller, L., Dillman, E., and Mackler, B. (1979). *J. Clin. Invest.* **64,** 129–137.
Finch, C. A., and Huebers, H. (1982). Perspectives in iron metabolism. *N. Engl. J. Med.* **306**(25), 1520–1528.
Finch, S., Haskins, D., and Finch, C. A. (1950). Iron metabolism. Hematopoiesis following phlebotomy. Iron as a limiting factor. *J. Clin. Invest.* **29,** 1078–1086.
Garby, L., Irnell, L., and Werner, I. (1969). Iron deficiency in women of fertile age in a Swedish community. 3. Estimation of prevalence based on response to iron supplementation. *Acta Med. Scand.* **185,** 113–117.
Hallberg, L. (1981). Bioavailability of dietary iron in man. *Annu. Rev. Nutr.* **1,** 123–147.
Hallberg, L. (1982). Iron nutrition and food iron fortification. *Semin. Hematol.* **19**(1), 31–41.
Hallberg, L., Hogdahl, A. M., and Nilsson, L. (1966). Menstrual blood loss—a population study. *Acta Obstet. Gynaecol. Scand.* **45,** 25–56.
Labbe, R. F., and Finch, C. A. (1980). Erythrocyte protoporphyrin: Application in the diagnosis of iron deficiency. *In* "Methods in Hematology: Iron" (J. D. Cook, ed.), pp. 44–58. Churchill-Livingstone, Edinburgh and London.
Langer, E. E., Haining, R. G., Labbe, R. F., Jacobs, P., Crosby, E. F., and Finch, C. A. (1972). Erythrocyte protoporphyrin. *Blood* **40,** 112–128.
Martinez-Torres, C., Leets, I., Cobeddu, L., Layrisse, M., Dillman, E., Johnson, D. G., Brengelmann, G. L., and Finch, C. A. (1984). Effect of exposure to low temperature on normal and iron deficient subjects. *Am. J. Physiol.* **246,** R380–R383.
Monsen, E. R., Critchlow, C. W., Finch, C. A., and Donohue, D. M. (1984). Iron balance in superdonors. *Transfusion* **23**(2), 221–225.
Pollitt, E., and Leibel, R, L., eds. (1982). "Iron Deficiency: Brain Biochemistry and Behavior." Raven Press, New York.

Roche, M., and Layrisse, M. (1966). The nature and causes of 'hookworm anemia.' *Am. J. Trop. Med. Hyg.* **15,** 1029–1102.

World Health Organization (WHO). (1968). Nutritional anaemias. Report of a WHO Scientific Group. *W.H.O. Tech. Rep. Ser.* **405.**

World Health Organization (WHO). (1975). Control of nutritional anemia with special reference to iron deficiency: Report of an IAEA/USAID/WHO joint meeting. *W.H.O. Tech. Rep. Ser* **580.**

Worwood, M. (1981). Serum ferritin. *In* ''Iron in Biochemistry and Medicine, II'' (A. Jacobs and M. Worwood, eds.), pp. 203–244. Academic Press, New York.

2

Factors Influencing the Efficacy of Iron Fortification and the Selection of Fortification Vehicles

LEIF HALLBERG

University of Göteborg
Department of Medicine II
Göteborg, Sweden

I. INTRODUCTION

The beneficial effects of an iron fortification program can be predicted from (1) the estimated *amount of extra iron absorbed* by the target groups and (2) its *effect on the iron balance* in these groups. These two main points will be discussed separately. Much new knowledge has been obtained in both these areas. As well, a new methodology to study food iron absorption has been developed, and as a result of this, knowledge has been obtained about various factors in the diet influencing iron absorption and the bioavailability of iron compounds used for iron fortification. Iron requirements and their variation in different population groups are also better known. Thus, it is possible today to make fairly good predictions of the beneficial effects expected from iron fortification (Hallberg, 1981a, 1982).

II. AMOUNT OF EXTRA IRON ABSORBED BY IRON FORTIFICATION

There are several factors influencing the amount of extra iron absorbed by iron fortification (Table I). The main ones are related to properties of (1) the iron

ISBN 0-12-177060-5

Table I
FACTORS INFLUENCING THE EFFICACY OF IRON FORTIFICATION

Amount of fortification iron consumed
Physicochemical properties of iron compound
Interaction between fortification iron and vehicle during storage and food preparation
Properties of meals containing the iron fortificant with respect to bioavailability of its iron
Iron status of target population

compound, (2) the diet, and (3) the subjects. These three factors will be discussed separately.

A. Iron Compound

Some of the iron compounds used for iron fortification such as ferrous sulfate are potentially fully available for absorption, implying that the iron may completely mix with the nonheme iron pool (see below). Other iron compounds used for fortification are not so fully available for absorption because the iron may be only partly soluble under the conditions prevailing in the gastrointestinal tract. The fraction of the iron that forms part of the nonheme iron pool determines the *relative bioavailability* of an iron compound.

The relative bioavailability of an iron compound may not necessarily be constant. Some factors that might influence the relative bioavailability are the acidity of the gastric contents, the presence of compounds in the meal facilitating the rate of dissolution of the fortification iron (such as different acids), chelating compounds, and the time the food remains in the stomach. Methods used in food processing such as drying (Lee and Clydesdale, 1980a), heat and pressure processing (Wood *et al.*, 1978), thermal processing (Lee and Clydesdale, 1981), or in food preparation such as the time and temperature of bread fermentation (Lee and Clydesdale, 1980b), may also influence the dissolution of the iron compound. Variations in the amounts of promoters of iron absorption, such as ascorbic acid and meat, or the amounts of inhibitors of iron absorption such as tannins and bran, may not only influence the bioavailability of the nonheme *food* iron but also the *relative* bioavailability of an added iron compound.

Actually little is known about the effects of various absorption factors, food processing, and food preparation on the relative bioavailability in humans of iron compounds presently used for fortification.

The relative bioavailability of an iron compound can be determined in humans using radioiron methodology. The iron compound used for iron fortification is homogeneously labeled by a radioiron isotope (e.g., ^{55}Fe). The labeled fortification iron is then mixed into food—for example, with flour in making iron-fortified bread. A trace amount of a soluble inorganic iron salt labeled with another radioiron tracer (e.g., ^{59}Fe) is added to the dough. This iron will completely mix with the nonheme iron pool. This doubly radioiron-labeled bread is served alone or with a meal to volunteers. Two weeks later a blood sample is drawn to determine the ratio of ^{55}Fe to ^{59}Fe that is absorbed in relation to the amounts given. The relative bioavailability, under these specific experimental conditions, is thus the ratio of absorption between the iron compound used for fortification and of the iron tracer that completely mixes with the nonheme iron pool.

Determinations of the relative bioavailability of iron compounds to be used in fortifying foods must be made in humans, since there may be significant species differences. However, as studies in humans are quite laborious and as a great number of studies must be made to study variations in bioavailability induced by food processing and storage, it is necessary to find simpler, less time-consuming, but equally reliable methods. The most widely used method at present is based on iron absorption measurements (or hemoglobin regeneration) in iron-depleted rats (Fritz *et al.*, 1970; Pla and Fritz, 1971; Rees and Monsen, 1973). Other more indirect methods to compare the relative bioavailability of the iron compounds are based on physical measurements such as particle size distribution, relative surface area, or solubility—especially the rate of dissolution (Björn-Rasmussen *et al.*, 1977). Unfortunately, very few studies have been reported in which identical iron compounds have been compared in humans, in animals, and in different *in vitro* systems. It is thus difficult today to translate results obtained *in vitro* or in animals to relative bioavailability in humans. It must be stressed, however, that even if more knowledge of nonhuman systems were available it would still be necessary to make a great number of studies in humans, for, as mentioned earlier, the relative bioavailability of an iron compound in humans may be markedly affected by the meal composition or by the way the food is prepared. Thus, a figure obtained for the relative bioavailability of an iron compound to humans studied in a continental breakfast may not be the same as in meals containing much ascorbic acid or meat.

For these reasons there is a need to standardize the methodology for studies on relative bioavailability of compounds used for iron fortification as well as a need to limit the number of iron compounds used, in order to obtain a reasonably solid basis of inowledge about each compound and its relative bioavailability in humans under different conditions.

B. Diet

The extra iron absorbed by iron fortification is determined not only by the *relative bioavailability* of the iron compound used, as discussed earlier (fraction of iron mixing with the nonheme iron pool), but also to a very large extent by the properties of the meals containing the iron fortification source, which influences the fraction of the nonheme iron that will be absorbed from this pool (Cook *et al.*, 1973; Layrisse *et al.*, 1973; Hallberg, 1974; Hallberg *et al.*, 1978; for a review, see Hallberg, 1981a).

There are two kinds of iron in the diet with respect to the mechanism of absorption—heme iron (derived from hemoglobin and myoglobin) and nonheme iron (derived mainly from cereals, vegetables, and fruits). With few exceptions fortification iron is of the nonheme type. All nonheme iron compounds in a meal form part of a common nonheme iron pool (Hallberg and Björn-Rasmussen, 1972; Cook *et al.*, 1972). The properties of this pool are determined by the solubility and dissociation of the iron compounds present in different foods and by a balance between various factors present in the meal that may enhance or inhibit the absorption of iron. The addition of ascorbic acid, for example, will increase the absorption of all nonheme iron compounds present in the meal (Hallberg, 1981b). Fortification iron that is potentially available will be integrated in the nonheme iron pool of the meal. The iron absorption from this pool will result from the net effect of the properties of the meal and the chemical properties of that part of the fortification iron that is dissolved (Hallberg, 1974). A relative bioavailability of 50% of the iron compound means that half of the fortification iron will be integrated in the nonheme iron pool and will become potentially available for absorption.

Several studies have shown that the bioavailability of nonheme iron in a meal can vary widely depending on the composition of the meal (Hallberg, 1981a). The bioavailability of that part of the fortification iron that is integrated in the nonheme iron pool will therefore also vary in the same way.

In Western countries a great deal of fortification iron is consumed in bread. If most of the bread is eaten with meals that promote iron bioavailability, the effect of iron fortification will be much more marked than if it is mainly consumed in meals with a low content of enhancing factors and/or a high content of inhibitors. If, for example, the bread is consumed with breakfast meals and tea is usually consumed, about five to six times less fortification iron will be absorbed than if coffee and orange juice are consumed (Rossander *et al.*, 1979).

Most of the nonheme iron present in a Western-type meal is fully available for absorption. The unavailable fraction of the native iron is very small and can usually be disregarded (Hallberg and Björn-Rasmussen, 1982). It consists, for example, of some ferritin–hemosiderin iron (Layrisse *et al.*, 1975; Martinez-Torres *et al.*, 1976) and some ferric hydroxides (Derman *et al.*, 1982). In developing countries, on the other hand, the diet may contain rather considerable amounts of relatively unavailable iron compounds as a result of contamination of foods with dust and soil iron (Hallberg *et al.*, 1983).

It is necessary to consider the presence of contamination iron in the estimation

Table II

EFFECTS OF IRON FORTIFICATION ON IRON-DEFICIENT MEN ON THREE DIFFERENT HYPOTHETICAL DIETS

1. In one population the daily iron intake is 10 mg in adult men who are borderline iron deficient. The daily iron requirements can be set to 1 mg in these men, and thus, 10% of the dietary iron can be expected to be absorbed. If the daily iron intake is increased by 4 mg, and assuming that the fraction of dietary iron that is absorbed is constant, then the iron absorption will be increased by *0.4 mg*.

2. In another population the iron intake in a similar group of men is 20 mg; the calculated iron absorption is thus 5% (e.g., the diet has a lower bioavailability). If the same amount of fortification iron is added (4 mg), the absorption increase will be only *0.2 mg* (5% of 4 mg).

3. If in Example 2, 50% of the iron intake of 20 mg comes from unavailable contamination iron, then the absorption from the remaining 10 mg in the nonheme iron pool will be 10% as in Example 1. The addition of 4 mg fortification iron will then increase the absorption by *0.4 mg*.

of the effects expected from an iron fortification program. This is illustrated in three hypothetical examples in Table II. Fortification iron (4 mg) in the form of a fully available iron salt (e.g., ferrous sulfate) is added to three types of diets. Only the nonheme iron is considered in the examples. To simplify the calculations, men with borderline iron deficiency are chosen as a target group.

C. Iron Status of Subjects

Iron absorption is inversely related to the amount of storage iron in the body (Bothwell *et al.*, 1979). The additional amount of iron absorbed by iron fortification will thus be influenced by the test subjects' iron status. The purpose of iron fortification is to prevent the development of iron deficiency. This implies that one would like to know how much more iron is absorbed by iron fortification in subjects who are borderline iron deficient (i.e., subjects with no iron stores but who have not yet developed any anemia). This information can be obtained in absorption studies in humans by relating, in each subject, the measured iron absorption increase from iron fortification to the absorption from a reference dose of ferrous ascorbate (3 mg Fe^{2+}) given in a fasting state. The absorption from this dose is directly related to the iron status, and as borderline iron-deficient subjects absorb 40% from such a dose, all food iron absorption measurements can be "normalized" to this value (Magnusson *et al.*, 1981).

III. EFFECT ON IRON BALANCE

Knowing (1) the amount of fortification iron that reaches the target group(s), (2) the relative bioavailability of the iron compound, and (3) the bioavailability of the nonheme iron in the meals containing the fortification iron, it is then

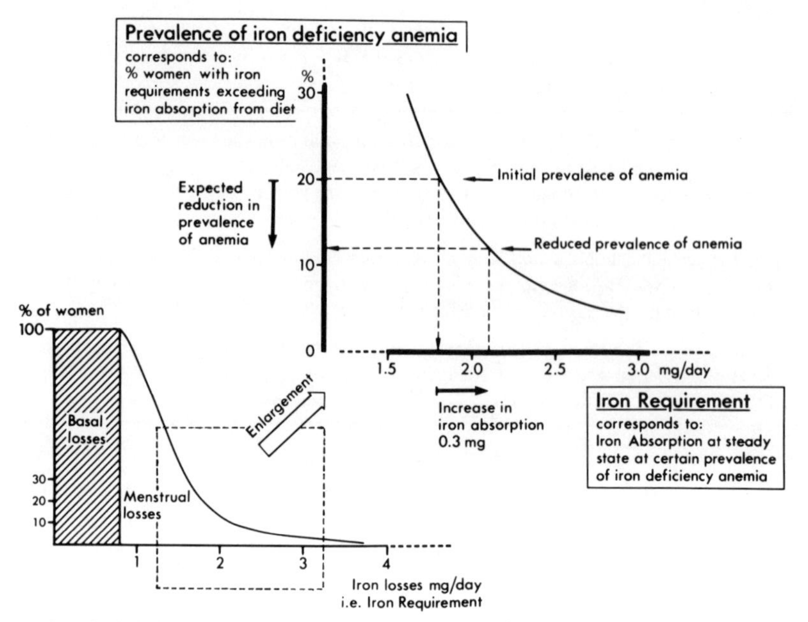

Figure 1. Cumulative distribution curve of iron requirements in menstruating women. The calculations are based on a body weight of approximately 55 kg, giving a basal iron loss of about 0.8 mg/day (lower left). An enlargement of the curve is shown in the upper right.

possible to calculate the amount of extra iron absorbed by the target group(s). To calculate the effect of this amount of absorbed iron on the iron balance in the target group, it is then necessary to know (1) the prevalence of iron deficiency in the target group and (2) the distribution of the iron requirements in the population at large. The rationale for such calculations has been published (Hallberg, 1982) and a simplified model is outlined in Fig. 1.

Figure 1 shows the cumulative distribution of the iron requirements in menstruating women. The ordinate represents the percentage of women with different iron losses and shows that 20% have an iron loss exceeding 1.8 mg daily. If the prevalence of iron deficiency is 20%, then this relationship implies that the diet can cover a daily iron loss of up to 1.8 mg. If iron fortification is calculated to lead to an absorption increase of 0.3 mg in the borderline subjects, then the prevalence of iron deficiency will fall to about 12%. If, instead, the initial prevalence had been only 10%, then the effect of the same absorption increase would be a reduction in prevalence to about 7%, reflecting a difference in the slope of the distribution curve of the iron requirements in these women.

In a male population infested with hookworms, the iron requirements may be distributed as shown in Fig. 2. With this form of the distribution curve, the effect of the same iron fortification program will be much smaller. Assuming that the

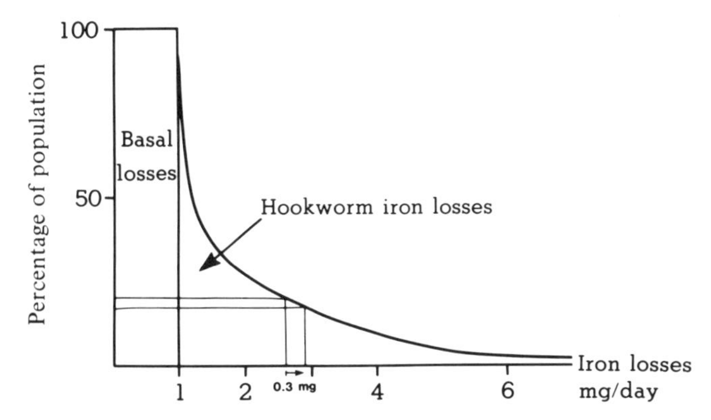

Figure 2. Cumulative distribution curve of iron requirements in a male rural population infested with hookworms. [Based on data from Layrisse and Roche (1964) and calculations by Baker and DeMaeyer (1979), and a basal iron loss of about 1 mg/day.]

initial prevalence of iron deficiency is 20% in this male population and that the absorption increase by iron fortification is 0.3 mg, the prevalence of iron deficiency can only be expected to decrease from 20 to about 18%. Such a small change will be impossible to detect in a reasonably large sample observed for a reasonable period of time.

These two examples show the importance of knowing the distribution of the iron requirements in a population before choosing a strategy for combatting iron deficiency in a population.

IV. CRITERIA FOR THE SELECTION OF IRON SOURCES

There are several factors to consider in choosing a fortifying compound (World Health Organization, 1975; Cook and Reusser, 1983). The following are examples of properties to keep in mind:

1. Relative bioavailability of the compound
2. Reactivity of the compound to cause discoloration or undesired changes in flavor or odor
3. Stability of the compound under storage and food preparation
4. Compatibility with other nutrients

The main problem in selecting an iron compound is that the better the relative bioavailability, the greater the probability that properties 2, 3, and 4 are poorly met. Conversely, the lower the relative bioavailability, the more iron needs to be given to achieve a certain effect, thus increasing costs and the probability of inducing undesirable effects. Obviously it is difficult to find the ideal iron compound for iron fortification.

Very few studies of the relative bioavailability of different iron compounds have been conducted in humans. Cook *et al.*, in 1973, reported studies in which ferrous sulfate was compared with sodium iron pyrophosphate, ferric orthophosphate, and reduced iron. Rolls were simultaneously tagged with ^{59}Fe as ferrous sulfate, and one of the three iron compounds was labeled with ^{55}Fe. The relative bioavailability of iron pyrophosphate was 0.02, ferric orthophosphate 0.35, and reduced iron 0.95. This latter iron was reduced by hydrogen and milled to small particles, most of them ranging in size from 5 to 10 μm. In our laboratory Björn-Rasmussen *et al.* (1977) studied four different preparations of hydrogen-reduced iron. By modifications of the manufacturing process, it was possible to prepare radioiron-labeled reduced iron with different solubility properties. The relative bioavailability varied from 1.00 to 0.13.

The rate of dissolution of different ferric orthophosphate compounds on the market varies more than 10-fold (L. Hallberg, personal observations), and it is probable that the relative bioavailabilities of the different compounds also vary in a similar manner. Since it is possible to prepare ferric orthophosphate with different solubility properties, we studied a batch of radioiron-labeled ferric orthophosphate that had an optimal rate of dissolution. Despite this rapid dissolution, the relative bioavailability was only 0.38 when studied in a Southeast Asian type of meal composed of rice, boiled vegetables, and a curry (L. Hallberg and E. Björn-Rasmussen, unpublished).

Cook *et al.* (1973) reported a similar relative bioavailability of ferric orthophosphate of 0.31. In this study the iron compound was baked into wheat rolls. Under similar conditions Cook *et al.* found the relative bioavailability of sodium iron pyrophosphate to be only 0.05. When the labeled iron pyrophosphate and a tracer of an iron salt labeled with another radioiron isotope were given in the same roll, which gives a more accurate figure of the relative bioavailability, the ratio was only 0.02.

Sodium iron ethylenediaminetetraacetic acid (FeNa EDTA) is a compound that has been much studied by several groups because of its high relative bioavailability (Layrisse *et al.*, 1976; Viteri *et al.*, 1978; Martinez-Torres *et al.*, 1979; MacPhail *et al.*, 1981). The relative advantage of FeNa EDTA seems to be greater in meals where the natural iron has a low bioavailability. As shown in Table III, the iron absorption is about two to three times higher from FeNa EDTA than from simple iron salts such as ferrous or ferric sulfate when added to maize meals. In rice meals and gruel composed of wheat and oats, this ratio is about 1.6 and 1.2, respectively. In a hamburger meal no significant difference is seen between FeNa EDTA and ferrous sulfate. These observations imply that the meal composition affects the relative bioavailability of FeNa EDTA and that this iron compound may be most useful as an iron fortificant in meals with a high content of factors inhibiting nonheme iron absorption.

Table III
COMPARISON OF ABSORPTION OF IRON FROM $FeSO_4$- OR FeNa EDTA-FORTIFIED MEALS

Number of subjects	Amount of iron added (mg)	Absorption (%)		A/B	Reference dose absorption (%)	Type of meal	Reference
		$FeSO_4$ (A)	FeNa EDTA (B)				
12	5.0	3.5	7.2	2.06	35.3	Maize porridge	MacPhail *et al.* (1981)
7	2.5	3.5	9.8	2.80	76.8	Milk, rice, sugar formula	Viteri *et al.* (1978)
21	5.0	3.3	7.4	2.20	51.3	Black bean gruel, corn tortillas, wheat bread, coffee	Viteri *et al.* (1978)
12	3.0	7.7	18.0	2.30	49.3	Wheat dough	Martinez-Torres *et al.* (1979)
11	3.0	12.4	14.4	1.20	35.2	Sweet manioc	Martinez-Torres *et al.* (1979)
18	5.0	3.9	6.3	1.62	43.0	Rice, boiled vegetables, curry (Thai meal)	L. Hallberg and L. Rossander (unpublished results)
8	2.7	6.6	7.9	1.20	47.8	Wheat–oat gruel	
10	5.0	5.4	5.7	1.05	35.0	Hamburger, string beans, potatoes	

Two other approaches to increase the iron absorption from a meal should be mentioned. Heme iron has been considered as a fortification agent. Dried slaughterhouse blood has been baked into bread under the assumption that the high bioavailability of heme iron when ingested in meat could be preserved. Unfortunately, however, the bioavailability of heme iron is much lower when not given together with meat (Hallberg *et al.*, 1979). Heme iron is probably an ideal compound with which to fortify meat products, for example, hamburgers in which part of the meat has been replaced with soy protein (Hallberg and Rossander, 1982).

Another method to increase the iron absorption from a meal is to improve the bioavailability of iron in the meal by giving compounds enhancing the absorption of iron, such as ascorbic acid. The value of ascorbic acid, however, is limited because of its destruction by oxidation and heating. Because of the marked absorption-promoting effects of ascorbic acid, it is a challenge to develop techniques of using this compound as an alternative or complement to iron fortification.

V. CRITERIA FOR SELECTING THE FORTIFICATION VEHICLE

Several factors need to be considered in the choice of vehicles for iron fortification:

1. The vehicle must be consumed in sufficient amounts by the target groups in the population. The variation in the consumption should not be too great between different subjects. Ideal vehicles in most countries from this point of view are salt and flour, where the variation in consumption is only about twofold.
2. The vehicle should be produced in only a few centers or distributed through only a few channels so that the fortification can be adequately monitored.
3. The fortified vehicle should remain stable and palatable after fortification; that is, the fortification should not change the properties of the vehicle. For example, if mixed into flour it should not interfere with the baking properties; it should not cause any unacceptable discoloration under generally prevailing conditions. Moreover, the usual food preparation technique of the vehicle should not impair the bioavailability of the iron.
4. The distribution of the fortificant in the vehicle should not change during storage; for instance, there should be no sedimentation of iron in the fortified flour. In several developing countries, there are a very limited number of foods that come from only a few production or distribution centers. Moreover, in such countries there is often a need to fortify the diet not only with iron but also with other nutrients such as iodine and vitamin A. Under such conditions there may be a competitive situation, especially as there is interaction between iron and other

items. When salt is used as a vehicle for iron fortification, and the iron compound used is an iron salt, it is necessary to keep the pH of the salt low to prevent discoloration and formation of poorly available ferric hydroxides. The bioavailability of iron from such acidified salts may then be good. Unfortunately, it is no longer possible to use the same salt as a vehicle for iodine fortification, as iodine will disappear by evaporation at low pH. If Fe EDTA is used instead of an iron salt to fortify salt, there is no need to keep a low pH and it is then possible to fortify with both iron and iodine.

This example is given to illustrate the sometimes complex interaction between iron, vehicle, and other fortificants.

Several vehicles have been used for iron fortification, including cereals, salt, infant foods, sugar, drinks, and condiments. The technical application will be described in Chapters 6 through 13.

REFERENCES

Baker, S. J., and DeMaeyer, E. M. (1979). Nutritional anemia: Its understanding and control with special reference to the work of the World Health Organization. *Am. J. Clin. Nutr.* **32,** 368–417.

Björn-Rasmussen, E., Hallberg, L., and Rossander, L. (1977). Absorption of fortification iron. Bioavailability in man of different samples of reduced iron and prediction of the effects of iron fortification. *Br. J. Nutr.* **37,** 375–388.

Bothwell, T. H., Charlton, R. W., Cook, J. D., and Finch, C. A. (1979). "Iron Metabolism in Man." Blackwell, Oxford.

Cook, J. D., and Reusser, M. E. (1983). Iron fortification: An update. *Am. J. Clin. Nutr.* **38,** 648–659.

Cook, J. D., Layrisse, M., Martinez-Torres, C., Walker, R., Monsen, E., and Finch, C. A. (1972). Food iron absorption measured by an extrinsic tag. *J. Clin. Invest.* **51,** 805–815.

Cook, J. D., Minnich, V., Moore, C. V., Rasmussen, A., Bradley, W. B., and Finch, C. A. (1973). Absorption of fortification iron in bread. *Am. J. Clin. Nutr.* **26,** 861–872.

Derman, D. P., Bothwell, T. H., Torrance, J. D., MacPhail, A. P., Bezwoda, W. R., Charlton, R. W., and Mayet, F. G. H. (1982). Iron absorption from ferritin and ferric hydroxide. *Scand. J. Haematol.* **29,** 18–24.

Fritz, J. C., Pla, G. W., Roberts, T., Boehne, J. W., and Hove, E. L. (1970). Biological availability in animals of iron from common dietary sources. *J. Agric. Food Chem.* **18,** 647–651.

Hallberg, L. (1974). The pool concept in food iron absorption and some of its implications. *Proc. Nutr. Soc.* **33,** 285–291.

Hallberg, L. (1981a). Bioavailability of dietary iron in man. *Annu. Rev. Nutr.* **1,** 123–147.

Hallberg, L. (1981b). Effect of vitamin C on the bioavailability of iron from food. *In* "Vitamin C Ascorbic Acid" (J. N. Counsell and D. H. Hornig, eds.), pp. 49–61. Appl. Sci. Publ., London.

Hallberg, L. (1982). Iron nutrition and food iron fortification. *Semin. Hematol.* **19,** 31–41.

Hallberg, L., and Björn-Rasmussen, E. (1972). Determination of iron absorption from whole diet. A new two-pool model using two radioiron isotopes given as haem and non-haem iron. *Scand. J. Haematol.* **9,** 193–197.

Hallberg, L., and Björn-Rasmussen, E. (1982). Measurement of iron absorption from meals contaminated with iron. *Am. J. Clin. Nutr.* **34,** 2808–2815.

Hallberg, L., and Rossander, L. (1982). Effect of soy protein on non-heme iron absorption in man. *Am. J. Clin. Nutr.* **36,** 514–520.

Hallberg, L., Björn-Rasmussen, E., Garby, L., Pleehachinda, R., and Suwanik, R. (1978). Iron absorption from South-East Asian diets and the effect of iron fortification. *Am. J. Clin. Nutr.* **31,** 1403–1408.

Hallberg, L., Björn-Rasmussen, E., Howard, L., and Rossander, L. (1979). Dietary heme iron absorption. A discussion of possible mechanisms for the absorption-promoting effect of meat and for the regulation of iron absorption. *Scand. J. Gastroenterol.* **14,** 769–779.

Hallberg, L., Björn-Rasmussen, E., Rossander, L., Suwanik, R., Pleehachinda, R., and Tuntawiroon, M. (1983). Iron absorption from some Asian meals containing contamination iron. *Am. J. Clin. Nutr.* **37,** 272–277.

Layrisse, M., and Roche, M. (1964). The relationship between anaemia and hookworm infection. *Am. J. Hyg.* **79,** 279–301.

Layrisse, M., Martinez-Torres, C., Cook, J. D., and Finch, C. A. (1973). Iron fortification of food: Its measurement by the extrinsic tag. *Blood* **41,** 333–352.

Layrisse, M., Martinez-Torres, C., Renzi, M., and Leets, I. (1975). Ferritin iron absorption in man. *Blood* **45,** 689–698.

Layrisse, M., Martinez-Torres, C., and Renzi, M. (1976). Sugar as a vehicle for iron fortification. Further studies. *Am. J. Clin. Nutr.* **29,** 274–279.

Lee, K., and Clydesdale, F. M. (1980a). Chemical changes of iron in food and drying processes. *J. Food Sci.* **45,** 711–715.

Lee, K., and Clydesdale, F. M. (1980b). Effect of baking on the forms of iron in iron-enriched flour. *J. Food Sci.* **45,** 1500–1504.

Lee, K., and Clydesdale, F. M. (1981). Effect of thermal processing on endogenous and added iron in canned spinach. *J. Food Sci.* **46,** 1064–1073.

MacPhail, A. P., Bothwell, T. H., Torrance, J. D., Derman, D. P., Bezwoda, W. R., Charlton, R. W., and Mayet, F. G. H. (1981). Factors affecting the absorption of iron from Fe(III)–EDTA. *Br. J. Nutr.* **45,** 215–227.

Magnusson, B., Björn-Rasmussen, E., Hallberg, L., and Rossander, L. (1981). Iron absorption in relation to iron status. Model proposed to express results of food iron absorption measurements. *Scand. J. Haematol.* **27,** 201–208.

Martinez-Torres, C., Renzi, M., and Layrisse, M. (1976). Iron absorption by humans from hemosiderin and ferritin. Further studies. *J. Nutr.* **106,** 128–135.

Martinez-Torres, C., Romano, E. L., Renzi, M., and Layrisse, M. (1979). Fe(III)–EDTA complex as iron fortification. Further studies. *Am. J. Clin. Nutr.* **32,** 809–816.

Pla, G. W., and Fritz, J. C. (1971). Collobarative study of the hemoglobin repletion test in chicks and rats for measuring availability of iron. *J. Assoc. Off. Anal. Chem.* **54,** 13–17.

Rees, J. M., and Monsen, E. R. (1973). Absorption of fortification iron by the rat: Comparison of type and level of iron incorporated into mixed grain cereal. *J. Agric. Food Chem.* **21,** 913–915.

Rossander, L., Hallberg, L,, and Björn-Rasmussen, E. (1979). Absorption of iron from breakfast meals. *Am. J. Clin. Nutr.* **32,** 2484–2489.

Viteri, F. E., Garcia-Hanez, R., and Torun, B. (1978). Sodium iron NaFeEDTA as an iron fortification compound in Central America. Absorption studies. *Am. J. Clin. Nutr.* **31,** 961–971,

Wood, R. J., Stake, P. E., Eiseman, J. H., Shippee, R. L., Wolski, K. E., and Koekn, U. (1978). Effects of heath and pressure processing on the relative biological value of selected dietary supplemental inorganic iron salts as determined by chick hemoglobin repletion assay. *J. Nutr.* **108,** 1477–1484.

World Health Organization (1975). Control of nutritional anaemia with special reference to iron deficiency. *W.H.O. Tech. Rep. Ser.* **580.**

II

TYPES OF IRON FORTIFICANTS

3

Elemental Sources

JOHN PATRICK, JR.
SCM Metal Products, Chemical Division
SCM Corporation
Johnstown, Pennsylvania

I. INTRODUCTION

Much of the early work done on evaluating elemental iron as a dietary iron source led to confusing bioavailability data, because investigators did not fully describe or characterize the iron used. If the iron source was not a salt or other compound, it was simply identified as reduced iron or ferrum reductum. During the past decade many of the studies have been repeated and much has been learned about the food iron source now called *elemental iron powder*.

There are three basic members of the elemental iron powder family, each one differing chemically and physically from the other because of its method of manufacture. As a further complication, changes in granulating and/or classifying individual members will also produce differences in bioavailability. Most producers of food grade iron powder are in the business of making all sorts of metal powders, and to them the need for chemical and physical identification is foremost. The intention here is to describe those characteristics of elemental iron powders, from a metallurgical viewpoint, that may be meaningful to both the food industry and the scientific community, as well as identifying the types known to be produced for this purpose.

II. REDUCED

Reduced iron is made by reduction of ground iron oxide with hydrogen or carbon monoxide at an elevated temperature. The purity of the product is dictated

ISBN 0-12-177060-5

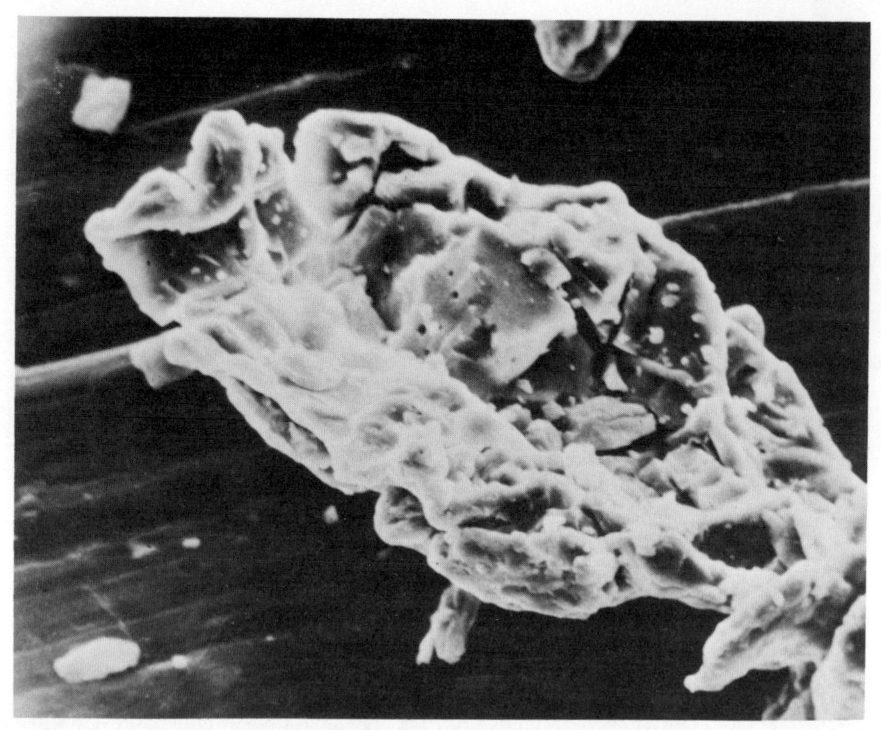

Figure 1. Electron micrograph (magnification 5000×) of a hydrogen-reduced iron particle.

by the purity of the iron oxide used. Most domestic reduced irons are made from mill scale rather than iron ore because the former is more pure. However, these products have the lowest purity of the food grade iron powders used. The basic impurities include C, Mg, Al, Si, P, S, Cr, Mn, Ni, and Cu, many of which are present as oxides and appear in the "acid insolubles" analysis. Because of the nature of the manufacturing process, iron mill scales reduced in carbon monoxide tend to have higher carbon (graphitic and/or combined) and sulfur. By far the greatest impurity in any iron powder is oxygen, and most of this occurs as a thin-film surface oxide that forms as the product is mechanically ground to powder (exposing more surface). In the case of reduced iron, there is in addition the probability that some of the original iron oxide will remain as a core in the center of the particles.

Reduced iron powder is a brittle material that lends itself well to comminution by ball milling, hammer milling, or attrition milling (Fig. 1). The particle shape is considered spongelike, irregular, and porous. The particle consists of a number of small, equiaxed grains, along whose boundaries can be found those impurity inclusions that are not alloyed with iron. These, together with iron oxide inclusions, provide weak points that contribute to its friability.

III. ELECTROLYTIC

Electrolytic iron powder is produced by electrolytic deposition of a hard, brittle metal that is mechanically comminuted. The iron is produced domestically using chemically pure iron anodes, a ferrous sulfate bath (electrolyte), and thin stainless steel cathode sheets onto which the iron migrates. These sheets are removed from the bath after a standard plating cycle, washed to remove soluble salts, dried, then flexed to remove the brittle deposit as fragments. These fragments are then mechanically ground to a finely divided powder. Insoluble contaminants originating from the anode fall to the bottom of the bath as a sludge, while electrochemical conditions are set to favor the migration primarily of iron ions to the cathode. Impurities that remain are usually present at levels of hundreds of parts per million or less. Again, as was described before, surface oxide is the major "impurity."

The particle shape of electrolytic iron powder is described as irregular, dendritic, or fernlike, from which it receives its high surface factor. Unlike reduced iron, the grains in electrolytic particles are less symmetrical (Fig. 2). Powders obtained by the electrolytic method are generally somewhat harder than those produced by reduction, so that grinding to powders with greater content of subsieve size is possible.

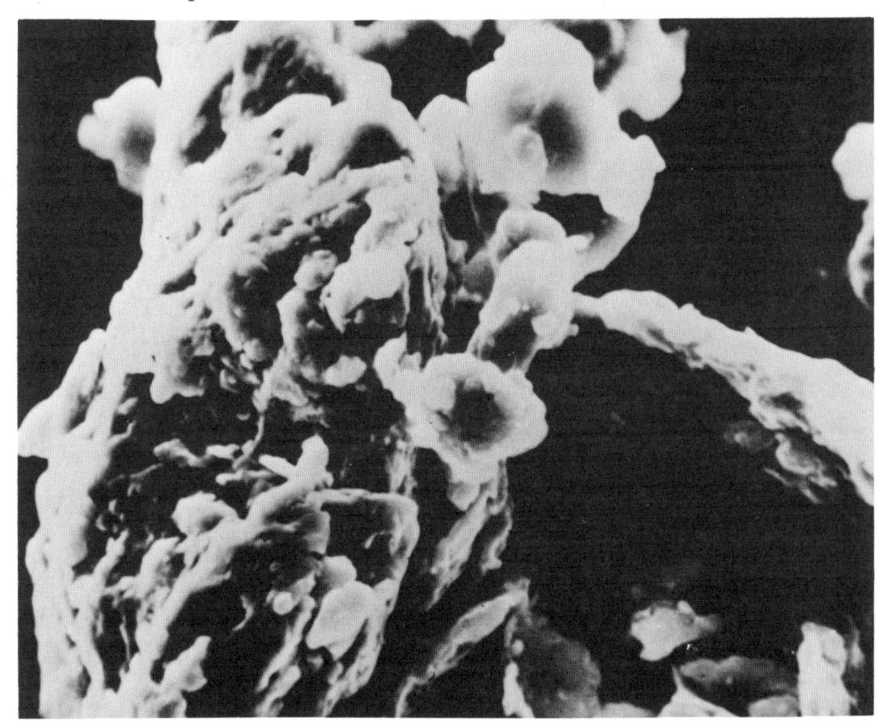

Figure 2. Electron micrograph (magnification 5000×) of an electrolytic iron particle.

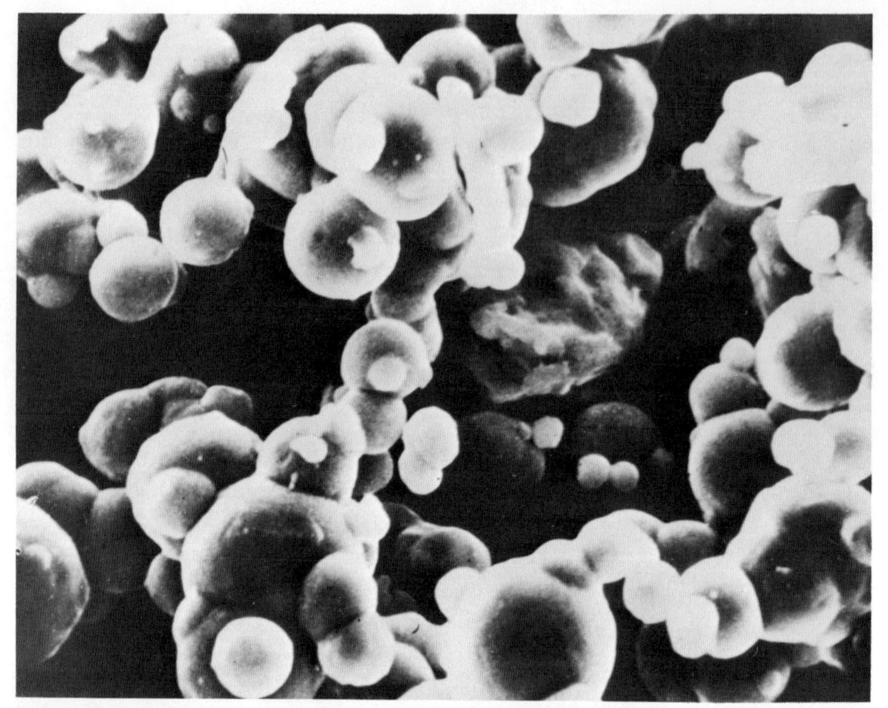

Figure 3. Electron micrograph (magnification 5000×) of carbonyl iron particles.

IV. CARBONYL

The extent to which carbonyl iron powder is produced in the United States for food enrichment purposes has not been established. Its production and use abroad have been reported. The unusual property of extremely fine particle size has resulted in considerable laboratory evaluation in this country in the past decade, presupposing its use accordingly. The method of manufacture involves the treatment of scrap iron or reduced iron with carbon monoxide under heat and pressure. The resulting iron pentacarbonyl, $Fe(CO)_5$, is later decomposed under controlled conditions, yielding an iron powder and carbon monoxide gas. At this point the major impurity is carbon (~1%), and a further reduction in wet hydrogen is necessary to remove most of it. The highly pure powder has particles ranging in size from 0.5 to 10 μm in diameter.

The carbonyl iron particle is very nearly spherical in shape and often is a cluster of several spheres "fused" together. The structure of the particle is characterized by concentric shells arranged in onion-skin fashion (Fig. 3). The particle is very dense, smooth, and hard skinned, not readily prone to surface oxidation (oxygen content is about one-fifth that of reduced iron or electrolytic iron). The major impurities are oxygen, carbon, and nitrogen, while lesser amounts of Si, Cr, Mn, and Ni are usually present at about the same level as in electrolytic iron. The carbonyl process is the most costly of those discussed here.

V. PROPERTIES OF IRON POWDERS

As can be seen, each of the three members of the elemental iron powder family is produced differently and thus exhibits its own special set of properties. With regard to oxygen, the major impurity in iron powder, milling and process conveying are usually done under an inert atmosphere; however, feedstock to this equipment contains trapped air, which then leads to particle surface oxidation despite this protection. With freshly milled and subsequent package-protected iron powder, the surface oxide is present as ferrosoferric, $FeO \cdot Fe_2O_3$. The ferrous oxide portion is very soluble in stomach acid and conveniently provides attack sites for particle dissolution; ferric oxide has little or no bioavailability.

A powder is generally considered to consist of discrete particles of dry material with a maximum dimension of 1 mm. While the use of coarse elemental iron powder in the food industry is known to exist, most usage has been restricted to minus 100 mesh (149 μm or smaller) and minus 325 mesh (44 μm or smaller) powders as presently specified under the three forms of elemental iron listed in the *Food Chemicals Codex,* Third Edition (1981). In recent years there has been much discussion regarding the use of finer iron powder for food enrichment, because increased solubility (and bioavailability) is associated with a more finely divided iron (and surface characteristics and purity are equally important). With a given member of the elemental iron powder family this is essentially true, but the theory does not exactly translate from member to member. For example, using two identically narrow particle size distributions of 20 to 26 μm, it has been demonstrated that electrolytic had twice the bioavailability of hydrogen-reduced iron by standard rat tests (Association of Official Analytical Chemists, 1975).

It is appropriate now to discuss those properties that relate to an iron powder's particle size and shape, and that will permit an investigator to identify his sample suitably. As has been shown, there are different shapes among the forms of iron powders discussed, and several methods for characterization may have to be used to get a comprehensive analysis of a given powder. In any discussion of particle physics the ideal model would be a true sphere, and the greater the departure from this shape, the more difficult it is to apply fundamental laws and to predict property measurements. It should be mentioned at this point that new techniques for property measurement are constantly being developed, but it is far beyond the scope of this chapter to account for even a portion of them. Instead, those basic measurements used by the powder metallurgy industry will be discussed.

Sieving is the best known and most widely used method of size analysis, because it is simple, fast, and relatively inexpensive. The method is described fully by the American Society for Testing of Materials (Philadelphia) (1976a). Both laboratory and production sieves are available covering a range of apertures from 0.3 to 0.0017 inch. In terms of U.S. sieves of most importance to the present discussion, it may be noted that the 100-mesh sieve has an aperture of 0.0059 inch and the 325-mesh sieve an aperture of 0.0017 inch. In making a comparison between different iron powders it is relatively easy to record the percentage of each that remains on top of the 100, the 200, and the 325 mesh, and that which passes through the 325-mesh sieve. Sieve "blinding" will occur when particles have partly passed through the mesh and are held by their maximum diameter in the aperture. Analytical reproducibility will be poor unless the sieves are periodically cleaned (ultrasonic cleaning appears to work best).

Measurement of particle sizes below 325 mesh (44 μm) will require what is known as "subsieve" methods. Probably the most common method involves the Fisher Subsieve Sizer (Fisher Scientific Co., Pittsburgh, Pennsylvania). The instrument was developed in the 1940s and has gained widespread use in the metal powder industry. Pressure loss of air caused by friction with particle surfaces is measured, with the packing of small particles offering more resistance to air flow than one of large particles. The test is quick, and the value (not absolute) obtained is called a "Fisher Number," which is reported as an average particle size in micrometers. The method is best suited for comparing different lots of the same commercial powder, and is described by the American Society for Testing of Materials (1982b).

Another subsieve measuring apparatus found in many powder metallurgy laboratories is the Roller Air Analyzer (American Instrument Co., Silver Spring, Maryland). This is an air elutriation method that is accurate for solid spheres and that becomes less accurate with departure from that shape. This separator works on the basis of Stokes' law as applied to the fall of spherical particles in an upward uniform flow of air. A selected range of small-sized particles (say 0–10 μm) is carried away by an applied air stream and collected in a thimble. The apparatus is then adjusted and the next larger particle size fraction (10–20 μm) is separated. This is repeated for the 20- to 30- and 30- to 40-μm fractions, and what remains behind in the original sample tube is the 40- to 44-μm "oversize." On a given lot of iron powder, only the minus 325-mesh portion can be analyzed by this method, so that a particle size analysis for a 100-mesh powder would be presented by showing a sieve analysis and also a Roller analysis (indicating the percentage distribution in the "subsieve" portion of that particular lot). Where a Fisher Number determination requires only a few minutes to complete, a fractionation by the Roller Method can take as long as 8 hours; however, the data are obviously more meaningful. This elutriation method is covered by the American Society for Testing of Materials (1976b).

The property of particle shape has special importance, since it affects surface area, apparent (bulk) density, permeability (as in the Fisher Number test), and flow characteristics of the powder. It has been found that the best characterization of shape will be an approximation, and for the necessity of simplicity and better interpretation it is best treated as a two-dimensional rather than a three-dimensional model. This can be justified for most powders, since from the major axis (the longest dimension) usually two axes have practically the same dimensions. The microscope is the most dependable tool for examination of particle shape, and photomicrographs can be made for the purpose of comparing lot with lot, or member with member in the iron powders. The value of estimating particle shape of different samples and relating this to bioavailability provides still another means for identifying experimental behavior.

Another property of interest is the apparent density of a powder population. This should not be confused with the true density or specific gravity of the material comprising the particles (e.g., 7.86 g/cm^3 for iron). Apparent density is defined as the weight of a mass of powder loosely heaped in a given volume and is expressed in grams per cubic centimeter. Most commercial metal powders used are minus 100 mesh and have a "natural" particle size distribution, which means that in a loose-packing mode, the voids between near-sized particles are filled with smaller particles with the net result of minimal porosity in the bed and a high apparent density (e.g., 2.50 g/cm^3). As the particle size distribution is narrowed—say a minus 325-mesh powder is now to be tested—there are no larger particles to create accommodations for the ultrafines (which represent tremendous surface area). Now large voids are formed by arching or bridging of adjacent particles, and the result is a low apparent density (e.g., 2.00 g/cm^3). With irregular-shaped particles like the reduced irons and electrolytic iron, bridging is more pronounced in the finer granulations than it would be with carbonyl iron. There are at least a dozen other factors that influence apparent density, but the intent here is to draw attention to another standardized, convenient, and reproducible test that is helpful in identifying iron powders. ASTM standards are provided for free-flowing metal powders (American Society for Testing of Materials, 1982a) and non-free-flowing metal powders (American Society for Testing of Materials, 1982c), which usually are the finer powders. The flowmeter funnel, density cup, and stand are available as a unit from Alcan Metal Powders, Inc. (Elizabeth, New Jersey).

VI. ADVANTAGES AND DISADVANTAGES

Perhaps the most important advantage of elemental iron powder over iron salts and compounds is the stability of the material. Properly stored product, in original containers, can be kept for as long as 2 years without property change. Once incorporated into a food product, this stability guards against off-flavor, odor, and caking, and minimizes catalysis of discoloration and rancidity.

All forms of elemental iron powder contain 96% or more total iron, so that enriching with these products allows the addition of iron without anion and/or cation excess baggage. The lower bulk volume favorably affects operations from enrichment pouches to warehouse storage.

Among the disadvantages of elemental iron powder, the dark gray color and higher density are more often discussed. Many say that the graying effect is pronounced in the enrichment mix and not detectable in the finished food product by the consumer. In earlier cases, when coarser elemental iron powders were more widely used, there were some reports that "specks" of iron could actually be seen in the finished product. The higher density of iron compared to the milled cereal grain can lead to segregation if incorporation is not carefully performed. Finer elemental iron, with lower apparent density, coupled with more irregular particle shape tends to minimize segregation.

The reported variability of iron absorption among the different forms of commercial elemental iron powders has led to some confusion. Researchers have reported bioavailability values from 10 to 90% that of ferrous sulfate, clearly indicating the need for more research.

An effort has been made here to acquaint the scientist and user of food grade elemental iron powders with basic methods for identification, and beyond that, to make them aware of the differences among commercially available powders. Finally, it should be mentioned that producers of food grade elemental iron powders are equipped to measure the properties discussed, and they report most of them in their lot analyses. It is important that the user take advantage of the data and other services offered by the manufacturer.

REFERENCES

American Society for Testing of Materials. (1976a). Specification for sieve analysis of granular metal powders, *Annu. Book ASTM Stand.*, B-214.

American Society for Testing of Materials. (1976b). Subsieve analysis of granular metal powders by air classification. *Annu. Book ASTM Stand.*, B-293.

American Society for Testing of Materials. (1982a). Test method for apparent density for free-flowing metal powders. *Annu. Book ASTM Stand.*, B-212.

American Society for Testing of Materials. (1982b). Test method for average particle size of powders of refractory metals by the Fisher Subsieve Sizer. *Annu. Book ASTM Stand.*, B-330.

American Society for Testing of Materials. (1982c). Test method for apparent density for non-free-flowing metal powders. *Annu. Book ASTM Stand.*, B-417.

Association of Official Analytical Chemists. (1975). "Official Methods of Analysis," 12th Ed., pp. 857–859. AOAC, Washington, D.C.

Food Chemicals Codex. (1981). "Food Chemicals Codex," 3rd Ed. National Academy Press, Washington, D. C.

4

Nonelemental Sources

RICHARD F. HURRELL
Nestlé Products Technical Assistance Co., Ltd.
Research Department
La Tour-de-Peilz, Switzerland

I. INTRODUCTION

When choosing an iron source to fortify a food product, one has to consider the influence of the added iron on the organoleptic properties of the product, whether the iron source is likely to be sufficiently well absorbed by the person consuming it, and—especially in government-sponsored intervention programs—the cost of fortification. A soluble iron source is preferable for addition to liquid products (infant formula, beverages) if sedimentation is to be avoided; however, in nonliquid products (cereals, infant foods) less soluble sources may also be envisaged. In general, those iron sources that are freely soluble in water (e.g., ferrous sulfate, ferrous gluconate, ferrous lactate, ferric saccharate, ferric ammonium citrate, soluble ferric pyrophosphate) have a high relative bioavailability, at least in rat assays. These soluble sources, however, may cause off-flavor problems due to their metallic taste or due to their catalysis of fat oxidation reactions, and they may form unacceptable colored complexes with

ISBN 0-12-177060-5

other food components such as polyphenolic substances and sulfur-containing compounds. The less soluble iron sources, on the other hand, cause fewer organoleptic problems, but they may have high (ferrous fumarate, ferrous succinate), medium (ferric pyrophosphate, elemental iron), or low relative bioavailability values (sodium iron pyrophosphate, ferric orthophosphate, large-particle-size reduced iron).

Table I
IRON SOURCES USED IN FOOD FORTIFICATIONS

Iron source	Approximate Fe content (%)	Average relative bioavailability		Relative cost[c]
		Rat[a]	Human[b]	
Freely water soluble				
Ferrous sulfate·$7H_2O$	20	100	100	1.0
Ferrous gluconate	12	97	89	5.1
Ferrous lactate	19	—	106	4.1
Ferric saccharate	3–35	92*	—	4.1
Ferric ammonium citrate	18	107	—	5.2
Ferrous ammonium sulfate	14	99	—	2.1
Ferric pyrophosphate plus sodium citrate	10	103*	—	6.3
Ferric choline citrate	14	102	—	11.0
Slowly soluble				
Dried ferrous sulfate	33	100	100	0.65
Ferric glycerophosphate	15	93	—	10.5
Ferric citrate	17	73	31	4.8
Ferric sulfate	22	83	34	1.1
Ferric pyrophosphate plus ammonium citrate	16	83*	—	3.6
Poorly soluble				
Ferrous fumarate	33	95	101	1.3
Ferrous succinate	35	119*	123	4.1
Ferrous tartrate	22	77	62	3.9
Ferrous citrate	24	76	74	3.9
Almost insoluble or insoluble				
Ferric pyrophosphate	25	45	—	2.3
Ferric orthophosphate	28	7–32	31	4.1
Sodium iron pyrophosphate	15	14	15	3.5
Reduced elemental iron	97	8–76	13–90	0.5–1.0[d]

[a]Hemoglobin repletion test, all values from Fritz *et al.* (1970) and Fritz and Pla (1972) except those marked with an asterisk, which are from R. F. Hurrell and D. E. Furniss (unpublished, 1983).
[b]Absorption measured using radioiron isotopes, all values from Brise and Hallberg (1962b) except ferric orthophosphate and sodium iron pyrophosphate (Cook *et al.*, 1973), and elemental iron (Björn-Rasmussen *et al.*, 1977).
[c]Approximate cost relative to ferrous sulfate·$7H_2O$ (DM 362/100 kg, February, 1984) for the same level of total iron. Calculation for ferric saccharate is for material containing 10% Fe.
[d]From Bothwell *et al.* (1979).

The major iron sources used in food fortification are listed in Table I. They are divided on the basis of solubility in water, not acid, into freely water soluble, slowly soluble, poorly soluble, and insoluble sources, and their iron content, relative cost, and relative bioavailability in rats and humans are also given. In rat studies, iron compounds have been compared to ferrous sulfate for their ability to regenerate hemoglobin in iron-deficient rats. The relative bioavailability values (RBV) ($FeSO_4 \cdot 7H_2O$ = 100) are taken from Fritz *et al.* (1970), Fritz and Pla (1972), and our own unpublished experiments. In human studies, iron absorption in iron-replete healthy humans has been measured using radioactive iron salts. The results from Brise and Hallberg (1962b) were obtained by feeding radioactive iron sources (30 mg Fe) directly to fasting subjects, whereas those from Cook *et al.* (1973) were obtained by feeding iron-fortified bread rolls.

The relative bioavailability of an iron source should be used with caution, as it is only an approximate guide to potential absorption. Meal composition, the physicochemical environment provided by the food, and the enhancers and inhibitors of iron absorption in a meal, together with the iron status of the person consuming it, largely govern the level of iron absorption (Clydesdale, 1983; Hallberg, 1981). Even ferrous sulfate is poorly absorbed from cereal products because of the potent inhibitors they contain, although its absorption may be increased severalfold by the addition of ascorbic acid, an absorption enhancer (Derman *et al.*, 1980). There is also much confusion and contradiction in the literature surrounding iron bioavailability. This is due to the different methodologies used (Van Campden, 1983), the difficulties in extrapolating from rat to human, the influence of processing and storage on the solubility and chemical form of iron (Hodson, 1970; Theuer *et al.*, 1971, 1973; Clydesdale, 1983), the influence of the physical characteristics of poorly soluble iron sources on their bioavailability (Harrison *et al.*, 1976; Shah *et al.*, 1977), the use of radioactive iron sources that are of different specification to the commercial sources, and the use of the extrinsic-tag technique to measure the bioavailability of poorly soluble iron sources that do not completely enter the common iron pool.

II. COMMERCIAL SOURCES

A. Ferrous Sulfate

Ferrous sulfate is the cheapest and most widely used iron source in food fortification. It comes in two forms: the heptahydrate and dried ferrous sulfate. Ferrous sulfate heptahydrate ($FeSO_4 \cdot 7H_2O$) forms pale, bluish green crystals containing about 20% iron that are freely soluble in water. It is odorless and has a slight saline taste. Upon warming to 56°C the compound loses three molecules of water to form the tetrahydrate, and further warming to 65°C forms the monohydrate, which is stable to 300°C. The dried ferrous sulfate available commercially ($FeSO_4 \cdot xH_2O$) is grayish white in color; it has a metallic and astringent taste, and is only slowly soluble in water. It consists mainly of the monohydrate but also contains some tetrahydrate. The iron content is about 33%.

The heptahydrate form is established as having a high relative bioavailability, and it has become the reference standard in almost all iron bioavailability studies. It has by definition a relative bioavailability value of 100, although in human studies Brise and Hallberg (1962b) found that from less than 5 to almost 70% of the initial dose may be absorbed depending simply on the iron status of the subject. Studies using anemic rats (Park *et al.*, 1983) have indicated that the bioavailability of iron from ferrous sulfate might be reduced during storage of the salt. This matter, however, was not clear-cut and needs further investigation.

Ferrous sulfate is the most commonly added iron source in both powder and liquid infant formulas. In 1982, all infant formulas in the United States were fortified with ferrous sulfate, as were some 20% of infant cereals (Monsen *et al.*, 1983). Rios *et al.* (1973) measured the absorption of radioactively labeled ferrous sulfate by infants fed fortified milk-based or soy-based formulas (12–17 mg of added iron/liter). They recorded a mean iron absorption of 3.9% (0.7–23.1%) for milk-based formulas and 5.4% (1.0–21.9%) for soy-based formulas. In infant cereals they recorded an absorption of 2.7% (0.4–12.1%). The absorption from both formula and cereals can be greatly increased by the addition of ascorbic acid. Derman *et al.* (1980) reported that the iron absorption by women from an infant formula containing 12.7 mg of added iron/liter as ferrous sulfate was increased from 7.2 to 20.8% by the addition of 80 mg/liter of ascorbic acid.

Ferrous sulfate may catalyze fat oxidation reactions and cause off-flavors in cereals during storage as described in Chapter 7 (Anderson *et al.*, 1976; Harrison *et al.*, 1976; Hurrell, 1984). It can, however, be used in bread and bakery products that are stored for only a short time (Cook and Ruesser, 1983), although a decrease in loaf quality and bread volume may result (Harrison *et al.*, 1976). The dry product causes far less off-flavor problems than the heptahydrate; however, they both caused unacceptable color changes in infant cereals when hot milk or hot water were added (Hurrell, 1984). Ferrous sulfate has also been found to discolor dehydrated mashed potato (Sapers *et al.*, 1974), to cause off-flavors in whole milk (Edmondson *et al.*, 1971) and skim milk (Borenstein, 1971), and to produce black discolorations when added to tea and coffee (Disler *et al.*, 1975). It could be added to sugar without any problems, although it rapidly produced off-colors in salt unless a "coordination" agent such as sodium hexametaphosphate was added (Zoller *et al.*, 1980).

B. Encapsulated Ferrous Sulfate

Encapsulation is a good method of reducing the catalytic effect of ferrous sulfate on fat oxidation. Durkee Industrial Food Group (Cleveland, Ohio) produces a range of ferrous sulfate products encapsulated with hydrogenated oils, zinc stearate, and maltodextrin. We have investigated their bioavailability in the

Table II
RELATIVE BIOAVAILABILITY OF ENCAPSULATED FERROUS SULFATE

Form of compound	Relative bioavailability[a]
Dried ferrous sulfate	102
Dried ferrous sulfate encapsulated with:	
Partially hydrogenated soybean oil (mp 66°C)	101, 115
Partially hydrogenated palm oil (mp 58°C)	95
Mono- and diglycerides (mp 53°C)	101, 116
Partially hydrogenated soybean and cottonseed oil (mp 52°C)	79
Maltodextrin	87
Zinc stearate (mp 121°C)	70[b]
Ethyl cellulose	133
Electrolytic iron (Glidden, United States)	40
Ferric pyrophosphate (Paul Lohman, Emmerthal, West Germany)	58

[a]As measured in the rat hemoglobin repletion test ($FeSO_4 \cdot 7H_2O$ = 100).
[b]Adapted from Zoller *et al.* (1980).

hemoglobin repletion test as well as the keeping quality of cereal products enriched with some of these products. The products tested contained 60% dried ferrous sulfate and 40% encapsule, and the melting point (mp) of the hydrogenated oils varied from 52 to 66°C.

Encapsulation with the hydrogenated oils or with maltodextrin had little influence on the bioavailability of ferrous sulfate (Table II). The products encapsulated with partially hydrogenated soybean oil, partially hydrogenated palm oil, and mono- and diglycerides performed particularly well. Electrolytic iron (Glidden) and ferric pyrophosphate were tested at the same time and were only half as well absorbed.

When dry-mixed with precooked wheat flour (10 mg Fe/100 g) and stored at 37°C for 3 months, the encapsulated products did not cause any off-flavor problems or excessive fat oxidation as measured by a taste panel and by pentane generation. They performed as well as electrolytic iron and the unfortified control. Color problems arose, however, when hot water or hot milk were added to make a pap. As with ferrous sulfate, various shades of gray and green appeared at temperatures above the melting points of the capsules (Hurrell, 1984). We have not been able to obtain the zinc stearate-coated material (melting point 121°C), but it has been described by Zoller *et al.* (1980) as a light-colored, tasteless, free-flowing powder with an RBV of 70 as measured in rats.

Ethylcellulose-encapsulated ferrous sulfate can be obtained from Eurand International (Milan, Italy). We have found this compound to have an RBV of 133 and to give only a slight color formation in cereals mixed with hot water or milk. There seems little point in encapsulating other sources of iron.

C. Ferrous Gluconate

Ferrous gluconate, $Fe[CH_2OH—(CHOH)_4—COO]_2·2H_2O$, is a yellowish gray to greenish yellow powder with a slight odor of caramel. It is soluble in water, contains around 12% Fe, and has been reported to have a similar bioavailability to ferrous sulfate in both rat and human assays (Table I). It is used to replace ferrous sulfate in the oral treatment of iron deficiency when side effects are a problem, but is not often used in food fortification. It could presumably be used in similar situations to ferrous sulfate, but it is about five times more expensive than the heptahydrate for an equivalent amount of available iron, and about eight times more expensive than the dried product (Table I). It has been used to fortify infant formula (Saarinen and Siimes, 1979), grape juice, and malt extract (Kaltwasser and Werner, 1983), but it rapidly promotes fat oxidation and off-flavor in stored cereals (Hurrell, 1984).

D. Ferrous Lactate

Ferrous lactate, $Fe(CH_3—CHOH—COO)_2·3H_2O$, is a yellow-green powder with a slight odor and sweet taste. It contains a minimum of 19% iron and is soluble in water. It is on the Codex advisory list (Codex Alimentarius Commission, 1981) of iron sources recommended for addition to infant formulas, and in human studies has shown a similar bioavailability to ferrous sulfate (Brise and Hallberg, 1962b). Anemic rats fed liquid soy-based or milk-based infant formulas likewise absorbed iron from ferrous lactate and ferrous sulfate equally well (Theuer *et al.*, 1971, 1973). It presumably has the same disadvantages as all soluble ferrous salts, and has been reported to give slight off-flavors in liquid whole milk or skim milk (Demott, 1971) and to cause a darkening in the color of chocolate milks (Douglas *et al.*, 1981).

E. Ferric Saccharate

This compound, also known as saccharated ferric oxide, comes as an amorphous reddish brown powder, soluble in water and containing 3, 10, 20, or 35% iron. It has been used to treat iron deficiency anemia both orally and intravenously (Wade, 1977). It is made from ferrous chloride, sodium carbonate, saccharose, and sodium hydroxide. It has no fixed formula but is reported to be a mixture of ferric oxide and saccharose. We have found that the 10, 20, and 35% Fe-containing ferric saccharate samples (from Dr. Paul Lohmann, Emmerthal, West Germany) give relative bioavailability values of 94, 102, and 79, respectively, in the hemoglobin repletion test with rats. In human studies, using the extrinsic-tag technique, Kaltwasser and Werner (1983) have reported results indicating that both normal and anemic subjects absorb iron equally well from ferric saccharate- or ferrous gluconate-fortified grape juice. As already mentioned, ferrous gluconate and ferrous sulfate are similarly absorbed by humans (Brise and Hallberg, 1962b).

In addition to grape juice, ferric saccharate has been used by European companies to fortify infant formulas, chocolate drink powders, and infant cereals. Despite its solubility, we have found that it causes no fat oxidation or off-flavor problems when fortified infant cereals are stored. The highest level of fortification tested was 18.5 mg Fe/100 g. It is rather dark in color, however, and may be unacceptable in light-colored products. Nevertheless, ferric saccharate appears to be an iron source of similar RBV to ferrous sulfate that can be used in cereals with little or no serious organoleptic problems, and also in liquid products.

F. Ferric Ammonium Citrate

Ferric ammonium citrate is a reddish brown granular powder that is very soluble in water. It has an indefinite stoichiometry and contains 16.5–18.5% Fe. Along with ferrous sulfate and elemental iron, it is the only other iron source permitted in England and Wales for the addition to flour. It has a similar RBV to ferrous sulfate in rats, but some concern has been raised about its bioavailability to humans (Cook and Ruesser, 1983), as poor absorption was reported when it was baked into bread (Callender and Warner, 1968) or added to wheat chappatis (Elwood *et al.*, 1970). However, iron added to cereals is always poorly absorbed, and the 2% absorption values reported by Elwood *et al.* (1970) for ferric ammonium citrate added to chappatis are little different from the 2.7% absorption reported by Rios *et al.* (1973) by infants fed ferrous sulfate-fortified infant cereal. Derman *et al.* (1980) likewise reported that ferric ammonium citrate and ferrous sulfate were both poorly absorbed (about 1%) by women fed fortified infant cereal; however, addition of ascorbic acid greatly increased the absorption. When the investigators added 50 mg of ascorbic acid to 30 g of cereal containing 6.6 mg of iron (5 mg from ferrous ammonium citrate), the Fe absorption increased from 0.8 to 10.3%.

Ferric ammonium citrate has also been used to fortify infant formula and fruit-flavored beverages (Cook and Ruesser, 1983) and has been reported to be organoleptically acceptable in fluid skim milk (Borenstein, 1971). It is about five times more expensive than ferrous sulfate heptahydrate for the equivalent amount of iron.

G. Ferrous Fumarate

Ferrous fumarate, Fe(COO—CH=CH—COO), is a reddish orange to reddish brown powder that is odorless and almost tasteless. It contains about 33% Fe and is only slightly soluble in water. It has given a similar bioavailability to ferrous sulfate in both rat and human studies (Table I). It has been used to fortify corn–soya–milk preparations used by the U.S. Agency for International Development in overseas assistance programs and is also reported to be compatible with cocoa powder and coffee, although it can produce unacceptable flavors in liquid milk (Lee and Clydesdale, 1979). According to Monsen *et al.* (1983), some 3% of cereals in the United States in 1982 were fortified with ferrous fumarate.

H. Ferrous Succinate

Ferrous succinate, Fe(COO—CH_2—CH_2—COO), is a pale brownish yellow amorphous powder, almost tasteless but with a slight odor. It contains 34–36% Fe in the dried form and is practically insoluble in water but soluble in dilute acids. When used to treat iron deficiency, it is claimed to cause fewer gastrointestinal side effects than ferrous sulfate. In both rat and human assays, it has given bioavailability values some 20% higher than ferrous sulfate (Table I). We have found no reports of its use as a food additive, although it is on the Codex advisory list of iron sources that may be used to fortify food for infants and young children (Codex Alimentarius Commission, 1981).

I. Ferric Pyrophosphate

Ferric pyrophosphate, $Fe_4(P_2O_7)_3 \cdot xH_2O$, contains 24–26% iron. It is a yellowish white, odorless powder, insoluble in water but soluble in mineral acids. It is widely used by European companies to fortify infant cereals and chocolate drink powders. It causes no organoleptic problems in cereals when added at around 10 mg Fe/100 g but at higher levels may provoke fat oxidation and off-flavors during storage. It is of mediocre bioavailability, giving an RBV of 45 to 60 in rat assays. The absorption of ferric pyrophosphate has never been satisfactorily measured in humans, partly because of the difficulty in obtaining a radioactive salt of the same specifications as the commercial product. Derman *et al.* (1980) attempted to measure the absorption of ferric pyrophosphate from an infant cereal by adding an extrinsic tag of radioactive ferric pyrophosphate, The absorption was similar to ferrous sulfate, but it is not clear whether the extrinsic tag and the fortification iron were of the same specifications.

J. Soluble Ferric Pyrophosphate and Orthophosphate

Ferric pyrophosphate may be solubilized with sodium citrate or ammonium citrate. These double salts are badly named and often confused with the insoluble ferric pyrophosphate used in cereal fortification in Europe and other parts of the world. Ferric pyrophosphate solubilized with sodium citrate (USP VIII) is an apple-green granular powder. It is odorless with an acidulous and slightly saline taste. It is soluble in water and contains a minimum of 10% Fe. We have found this material (from Dr. Paul Lohman, Emmerthal, West Germany) to have an RBV of 103 in the hemoglobin repletion test. Pla *et al.* (1973) have similarly reported an RBV of 96 for sodium citrate-stabilized ferric pyrophosphate. Ferric orthophosphate may be solubilized in a similar way and contains 12–15% Fe.

Ferric pyrophosphate solubilized with ammonium citrate gives yellowish green granules containing about 16% Fe that are slowly soluble in water. We have obtained an RBV of 83 for a sample purchased from Dr. Paul Lohmann (Emmerthal, West Germany). The sodium citrate-solubilized ferric pyrophosphate and orthophosphate were reported to be satisfactory for fortifying whole milk or skim milk (Demott, 1971; Borenstein, 1971), although we have found them to give color problems when hot water or hot milk are added to fortified infant cereals.

K. Ferric Orthophosphate

Ferric orthophosphate, $FePO_4 \cdot xH_2O$, is a yellowish to buff-colored powder, practically insoluble in water but soluble in dilute mineral acids. It is hydrated with two or four molecules of water and contains 26–30% Fe. Like elemental iron, its bioavailability appears to vary considerably from batch to batch in both rats and humans (Hallberg, 1981). Fritz *et al.* (1970) evaluated four samples in the hemoglobin repletion test with rats and reported RBVs from 7 to 32. Harrison *et al.* (1976) similarly found that the RBVs of five commercial samples of ferric orthophosphate varied from 6 to 46 (Table III), and they also demonstrated that it was strongly influenced by particle size and solubility in 0.1 *N* HCl. In a study

Table III
PHYSICAL CHARACTERISTICS AND BIOAVAILABILITY OF FERRIC ORTHOPHOSPHATE[a]

Particle size (μm)	Solubility in 0.1 *N* HCl, 3 hours (%)	RBV[b]
15	11.6	6
12	11.6	7
1	41.9	33
<1	45.5	33
<1	63.4	46

[a]Adapted from Harrison *et al.* (1976).
[b]Relative bioavailability value ($FeSO_4$ = 100).

with infants fed fortified infant cereal, Rios *et al.* (1973) found an RBV of 26, which was similar to the value of 31 found by Cook *et al.* (1973) when feeding adults the same iron source baked into bread rolls. Fairweather-Tait *et al.* (1983) have reported that ferric orthophosphate added to a cocoa drink was as well absorbed by human subjects as ferrous sulfate. However, they used the extrinsic-tag technique with the stable isotope ^{58}Fe, and it is not certain whether *in vivo* there was a complete isotopic exchange with the fortification iron.

In studies on the fortification of common salt in India, a combination of ferric orthophosphate and sodium hydrogen sulfate (absorption enhancer) has been reported to give acceptable long-term bioavailability of added iron with only slight discoloration even under adverse storage conditions (Nadiger *et al.*, 1980).

L. Sodium Iron Pyrophosphate

Sodium iron pyrophosphate, $Fe_4Na_8(P_2O_7)_5 \cdot xH_2O$, is a yellowish to tan powder containing 14.5–16% Fe. It is insoluble in water but soluble in hydrochloric acid. It has shown a variable but always low relative bioavailability in both rats and humans (Table I). In 1972, 50% of the infant cereals in the United States were fortified with this iron source, but by 1982 it was no longer used in infant cereal fortification (Monsen *et al.*, 1983).

M. Less Utilized Commercial Compounds

The following compounds are sometimes mentioned in the literature, but there is little evidence for their widespread use in food fortification.

1. FERROUS CITRATE

A gray powder containing approximately 24% Fe, ferrous citrate is a compound of indefinite stoichiometry; several forms of the salt are known, including the monohydrate and the decahydrate. It is practically insoluble in water and has given an RBV of 76 in rat assays and 74 in human studies (Table I).

2. FERRIC CITRATE

Ferric citrate is a brown granular powder containing 16.5–18.5% Fe that is slowly soluble in water. It is a combination of iron and citric acid of indefinite composition. In human studies, it has given an RBV of only 31; in rat studies a value of 73 has been reported (Table I). Both ferrous and ferric citrate are suggested by Codex as potential iron sources for addition to infant formula (Codex Alimentarius Commission, 1981). When added to liquid milk- or soy-based formula and heat-processed, the salts had the same RBV as added ferrous sulfate in the hemoglobin repletion test with rats (Theuer *et al.*, 1971, 1973).

3. FERRIC GLYCEROPHOSPHATE

Ferric glycerophosphate is a yellow to greenish yellow powder, $Fe_3[C_3H_5(OH)_2 \cdot PO_4]_3$, usually containing some water. It contains 13–16% Fe and is odorless, almost tasteless, and slowly soluble in water. It has given RBVs similar to ferrous sulfate in rat assays either alone (Fritz *et al.*, 1970) or when processed into liquid milk- or soy-based formula (Theuer *et al.*, 1971, 1973). There have been no human studies. It is about 10 times more expensive than ferrous sulfate heptahydrate for the equivalent amount of iron (Table I).

4. FERROUS AMMONIUM SULFATE

Ferrous ammonium sulfate, $Fe(NH_4)_2(SO_4)_2 \cdot 6H_2O$, consists of pale greenish blue crystals that are soluble in water and contains about 14% Fe. The RBV is equivalent to ferrous sulfate in rat assays (Table I).

5. FERRIC CHOLINE CITRATE

A greenish yellow brown powder, ferric choline citrate is soluble in water and contains approximately 14% Fe. Its RBV value in rat assays was 102, but it is 11 times as expensive as ferrous sulfate heptahydrate for the equivalent amount of iron (Table I).

6. FERROUS TARTRATE

Ferrous tartrate, Fe(COO—CHOH—CHOH—COO)$\cdot xH_2O$, a greenish powder, is slightly soluble in water and contains 21–23% iron. It has given an RBV of 77 in rat assays and 62 in human studies (Table I).

7. FERRIC SULFATE

Ferric sulfate, $Fe_2(SO_4)_3$, is a yellowish brown powder that is strongly hygroscopic. The commercial product normally contains about 20% water. It is only slowly soluble in water, however, and has given RBV values of 83 (65–100) in rat assays and 34 in human studies (Table I).

8. FERRIC CHLORIDE

Ferric chloride, $FeCl_3$, is very hygroscopic and tends to absorb water to form the hexahydrate, which comes as brownish yellow crystals. It is described by Yetley and Glinsmann (1983) as an "unpublished GRAS substance" of "adequate" bioavailability, although in animal assays it has given RBVs of 26 to 67, which could only be described as mediocre.

III. CONCLUSIONS

Ferrous sulfate is clearly the iron source of choice for the fortifcation of food products where applicable. It has a high relative bioavailability and is the least expensive iron fortificant. Other, more expensive soluble iron sources are of similar bioavailability but cause the same organoleptic problems as ferrous sulfate. The importance of price depends on the cost of the food being fortified. In infant formulas and infant cereals it is of minor importance, as the iron source counts for less than 0.1% of the total product cost. Fortification of common salt with iron in India, however, increased its cost by 20% (Cook and Ruesser, 1983).

When ferrous sulfate or other freely soluble salts cannot be used because of organoleptic problems, the food manufacturer has several alternatives (Table I). In general, the less water-soluble iron sources cause fewer organoleptic problems but have lower relative bioavailabilities. However, it is the solubility of the iron source in gastric juice that determines iron absorption in part, and some poorly water-soluble sources, such as ferrous fumarate and ferrous succinate, are highly soluble in dilute acid and have a high relative bioavailability. The insoluble elemental iron and insoluble phosphates are probably the most inert iron sources.

A soluble iron source is necessary for milk products and beverages, but other sources can be used if necessary for nonliquid products. The use of ferric orthophosphate and sodium iron pyrophosphate is not recommended because of their low and variable bioavailability. In cereals in the United States, these iron sources have been largely replaced with elemental iron, which is inert and has medium bioavailability (Cook and Ruesser, 1983). Elemental iron requires strict control, because—as with all iron sources—bioavailability may vary due to different physical and chemical characteristics of the various sources and different methods of production (Björn-Rasmussen *et al.*, 1977; Hurrell, 1984; Clydesdale 1983). Higher relative bioavailabilities of hydrogen-reduced iron may be achieved with particle sizes smaller than most commercial iron powders (Rios *et al.*, 1973; Cook *et al.*, 1973). However, the segregation of these high-density powders may be a problem in some foodstuffs, and finely powdered iron has been reported to oxidize on storage and to be pyrophoric (Lee and Clydesdale, 1979).

There may be better alternatives to elemental iron for the fortification of nonliquid products without necessarily reducing their shelf life. Ferric pyrophosphate has been widely used to fortify infant cereals in Europe, but this too is only of medium bioavailability and it provokes fat oxidation at higher levels of addition (>10 mg Fe/100 g). Other alternatives that deserve consideration are ferrous fumarate, ferrous succinate, and ferric saccharate, which are of equivalent bioavailability to ferrous sulfate, and ferrous tartrate and ferrous citrate, which are slightly less well absorbed. With the exception of ferric saccharate, all are poorly soluble in water and should cause less organoleptic problems. There is some indication that humans (but not rats) absorb ferric salts less well than ferrous salts (Lee and Clydesdale, 1979), and this would appear to be so for ferric citrate and ferric sulfate (Table I). To what extent it is simply due to solubility (Brise and Hallberg, 1962a) is not clear, as the bioavailability of soluble ferric salts has never been adequately measured in humans.

Irrespective of which iron source is finally chosen to fortify a particular foodstuff, the absorption enhancers and inhibitors present in that food and in the meal as a whole, together with the iron status of the person consuming the meal, will have a profound influence on iron absorption. Ascorbic acid is the most potent enhancer of iron absorption known, and it has been shown to increase severalfold the nonheme iron absorption from products such as cereals (Layrisse *et al.*, 1974) and soya (Morck *et al.*, 1982), which contain absorption inhibitors. It appears to increase the bioavailability of all iron salts (Derman *et al.*, 1980). Iron salts of medium relative bioavailability (RBV about 50), such as elemental iron and ferric pyrophosphate, could have increased absorption when consumed with sufficient ascorbic acid. Unfortunately, ascorbic acid is instable to heat, and its oxidation is catalyzed by iron (Borenstein, 1971). Losses may largely be overcome, however, by overaddition by the manufacturer, proper encapsulation (International Nutritional Anemia Consultative Group, 1982), careful storage, and careful preparation before consumption. Because of its well-proven enhancement of iron absorption, ascorbic acid should be added where possible, together with the chosen iron fortificant.

REFERENCES

Anderson, R. A., Vojnovich, C., and Bookwalter, G. N. (1976). Iron enrichment of dry-milled corn products. *Cereal Chem.* **53,** 937–946.

Björn-Rasmussen, E., Hallberg, L., and Rossander, L. (1977). Absorption of fortification iron. Bioavailability in man of different samples of reduced iron, and prediction of the effects of iron fortification. *Br. J. Nutr.* **37,** 375–388.

Borenstein, B. (1971). Rationale of food fortification with vitamins, minerals and amino acids. *CRC Crit. Rev. Food Technol.* **2,** 171–186.

Bothwell, T. H., Charton, R. W., Cook, J. D., and Finch, C. A. (1979). "Iron Metabolism in Man." Blackwell, Oxford.

Brise, H., and Hallberg, L. (1962a). A method for comparative studies on iron absorption in man using two radioiron isotopes. *Acta Med. Scand.* **171,** Suppl. 376, 7–22.

Brise, H., and Hallberg, L. (1962b). Absorbability of different iron compounds. *Acta Med. Scand.* **171,** Suppl. 376, 23–38.

Callender, S. T., and Warner, G. T. (1968). Iron absorption from bread. *Am. J. Clin. Nutr.* **21,** 1170–1174.

Clydesdale, F. M. (1983). Pysicochemical determinants of iron bioavailability *Food Technol.* **37,** 133–138.

Codex Alimentarius Commission (1981). "Report of Codex Committee on Foods for Special Dietary Uses," Alinorm 81/26. Joint FAO/WHO Food Standards Programme, FAO Rome.

Cook, J. D., and Ruesser, M. E. (1983). Iron fortification: An update. *Am. J. Clin. Nutr.* **38,** 648–659.

Cook, J. D., Minnich, V., Moore, C. V., Bradley, W. B., and Finch, C. A. (1973). Absorption of fortification iron in bread. *Am. J. Clin. Nutr.* **26,** 861–872.

Demott, B. J. (1971). Effects on flavor of fortifying milk with iron and absorption of iron from intestinal tract of rats. *J. Dairy Sci.* **54,** 1609–1614.

Derman, D. P., Bothwell, T. H., MacPhail, A. P., Torrance, J. D., Bezwoda, W. R., Charlton, R. W., and Mayett, F. G. H. (1980). Importance of ascorbic acid in the absorption of iron from infant foods. *Scand. J. Haematol.* **25,** 193–201.

Disler, P. B., Lynch, S. R., Charlton, R. W., Bothwell, T. H., Walker, R. B., and Mayet, F. G. H. (1975). Studies on the fortification of cane sugar with iron and ascorbic acid. *Br. J. Nutr.* **34,** 141–152.

Douglas, F. W., Rainey, N. H., Wong, N. P., Edmondson, L. F., and La Croix, D. E. (1981). Color, flavor and iron bioavailability in iron-fortified chocolate milk. *J. Dairy Sci.* **64,** 1785–1793.

Edmondson, L. F., Douglas, F. W., and Avants, J. K. (1971). Enrichment of pasteurized whole milk with iron. *J. Dairy Sci.* **54,** 1422–1426.

Elwood, P. C., Benjamin, I. T., Fry, F. A., Eakins, J. D., Brown, D. A., DeKock, P. C., and Shah, J. U. (1970). Absorption of iron from chapatti made from wheat flour. *Am. J. Clin. Nutr.* **23,** 1267–1271.

Fairweather-Tait, S. J., Minski, M. J., and Richardson, D. P. (1983). Iron absorption from a malted cocoa drink fortified with ferric orthophosphate using the stable isotope ^{58}Fe as an extrinsic label. *Br. J. Nutr.* **50,** 51–60.

Fritz, J. C., and Pla, G. W. (1972). Application of the animal haemoglobin repletion test to measurement of iron bioavailability in foods. *J. Assoc. Off. Anal. Chem.* **55,** 1128–1132.

Fritz, J. C., Pla, G. W., Roberts, T., Boehne, J. W., and Hove, E. L. (1970). Biological availability in animals of iron from common dietary sources. *J. Agric. Food Chem.* **18,** 647–651.

Hallberg, L. (1981). Bioavailability of dietary iron in man. *Annu. Rev. Nutr.* **1,** 123–147.

Harrison, B. N., Pla, G. W., Clarke, G. A., and Fritz, J. C. (1976). Selection of iron sources for cereal enrichment. *Cereal Chem.* **53,** 78–84.

Hodson, A. Z. (1970). Conversion of ferric to ferrous iron in weight control dietaries. *J. Agric. Food Chem.* **18,** 946–947.

Hurrell, R. F. (1984). Bioavailability of different iron compounds used to fortify formula and cereals: Technological problems. *In* "Iron Nutrition in Infancy and Childhood" (A. Steckel, ed.), pp. 147–178. Nestlé, Vevey/Raven Press, New York.

International Nutritional Anemia Consultative Group (INACG) (1982). "Iron Absorption from Cereals and Legumes." Nutrition Foundation, New York.

Kaltwasser, J. P., and Werner, E. (1983). Bioavailability of iron enriched dietetic foods. *In* "Groups with High Risk of Iron Deficiency in Industrialized Countries" (H. Dupin and S. Hercberg, eds.), pp. 309–320. INSERM, Paris.

Layrisse, M., Martinez-Torres, C., and Gonzalez, M. (1974). Measurement of total daily iron absorption by the extrinsic tag model. *Am. J. Clin. Nutr.* **27,** 152–162.

Lee, K., and Clydesdale, F. M. (1979). Iron sources used in food fortification and their changes due to food processing. *CRC Crit. Rev. Food Sci. Nutr.* **11,** 117–153.

Monsen, E. R., Rees, J. M., and Merril, J. (1983). Iron fortification of foods: United States 1972–1982. *In* "Groups with High Risk of Iron Deficiency in Industrialized Countries" (H. Dupin and S. Hercberg, eds.), pp. 321–329. INSERM, Paris.

Morck, T. A., Lynch, S. R., and Cook, J. D. (1982). Reduction of soy-induced inhibition of nonheme iron absorption. *Am. J. Clin. Nutr.* **36,** 219–228.

Nadiger, H. A., Krishnamachari, K. A. V. R., Nadamuni, N., Rao, B. S. N., and Srikantia, S. G. (1980). The use of common salt (sodium chloride) fortified with iron to control anaemia: Results of a preliminary study. *Br. J. Nutr.* **43,** 45–51.

Park, Y. W., Mahoney, A. W., and Hendricks, D. G. (1983). Bioavailability of different sources of ferrous sulphate fed to anemic rats. *J. Nutr.* **113,** 2223–2228.

Pla, G. W., Harrison, B. N., and Fritz, J. C. (1973). Comparison of chicks and rats as test animals for studying the bioavailability of iron. *J. Assoc. Off. Anal. Chem.* **56,** 1369–1372.

Rios, E., Hunter, R. E., Cook, J, D., Smith, N. J., and Finch, C. A. (1973). The absorption of iron as supplements in infant cereal and infant formula. *Pediatrics* **55,** 686–693.

Saarinen, U. M., and Siimes, M. A. (1979). Iron absorption from breast milk, cow's milk and iron-supplemented formula: And opportunistic use of changes in total body iron determined by haemoglobin, ferritin and body weight in 132 infants. *Pediatr. Res.* **13,** 143–147.

Sapers, G. M., Panasiuk, O., Jones, S. B., Kalan, E. B., and Talley, F. B. (1974). Iron fortification of dehydrated mashed potatoes. *J. Food Sci.* **39,** 552–558.

Shah, B. G., Giroux, A., and Belonje, B. (1977). Specifications for reduced iron as a food additive. *J. Agric. Food Chem.* **27,** 845–847.

Theuer, R. C., Kemmerer, K. S., Martin, W. H., Zoumas, B. L., and Sarret, H. P. (1971). Effect of processing on availability of iron salts in liquid infant formula products: Experimental soy isolate formulas. *J. Agric. Food Chem.* **55,** 555–558.

Theuer, R. C., Martin, W. H., Wallender, J. F., and Sarret, H. P. (1973). Effect of processing on availability of iron salts in liquid infant formula products: Experimental milk-based formulas. *J. Agric. Food Chem.* **21,** 482–485.

Van Campden, D. (1983). Iron bioavailability techniques: An overview. *Food Technol.* **37,** 127–131.

Wade, A., ed. (1977). "Martindale. The Extra Pharmocopea," 27th Ed. Pharmaceutical Press, London.

Yetley, E. A., and Glinsmann, W. H. (1983). Regulatory issues regarding iron bioavailability. *Food Technol.* **37,** 121–126.

Zoller, J. M., Wolinsky, I., Paden, C. A., Hoskin, J. C., Lewis, K. C., Lineback, D. R., and McCarthy, R. D. (1980). Fortification of non-staple food items with iron. *Food Technol.* **34,** 39–47.

5
Experimental Fortificants

PATRICK MACPHAIL
ROBERT CHARLTON
THOMAS H. BOTHWELL
WERNER BEZWODA
Joint University/M.R.C. Iron and Red Cell Metabolism Unit
Department of Medicine
University of the Witwatersrand Medical School
Johannesburg, South Africa

I. INTRODUCTION

There are two forms of dietary iron: heme and nonheme. The heme iron is well absorbed, whatever the composition of the diet, while the bioavailability of nonheme iron is profoundly influenced by the composition of the diet. Substances that promote nonheme iron absorption include meat and a number of organic acids, the most important of which is ascorbic acid; inhibitory substances include polyphenols, phytate, and unidentified components in fiber (Cook, 1977; Bothwell *et al.*, 1979; Reinhold *et al.*, 1981; Fernandez and Phillips, 1982; Gillooly *et al.*, 1983). These observations have important implications in relation to iron fortification, since they explain why iron deficiency is so common in populations subsisting on predominantly cereal-based diets. Not only do these cereals contain inhibitors of iron absorption, but such diets also usually lack fruit and meat. Fortification of these diets with iron is of negligible benefit, since the substances that inhibit the absorption of the intrinsic iron present in the diet also prevent the absorption of the fortification iron (Bothwell *et al.*, 1979). For this

ISBN 0-12-177060-5

reason, several strategies have been employed to increase the bioavailability of fortification iron. Substances that have been used beside ascorbic acid (Sayers *et al.*, 1973), are sodium acid sulfate and sodium hexametaphosphate (Narasinga Rao and Vijayasarathy, 1975). Alternatively, it is possible to use as fortificants iron compounds that are not susceptible to the effects of inhibitory ligands present in the diet. The two that have been studied are the iron chelate, sodium ferric ethylenediaminetetraacetic acid [Fe(III)EDTA], and hemoglobin. This chapter reviews the work that has been done on these two substances and discusses their possible roles as iron fortificants.

II. SODIUM FE(III)EDTA

Fe(III)EDTA, a pale yellow powder that is soluble in water and has a high stability constant, has been used successfully for some years as an iron source in plants (Stewart and Leonard, 1952). In humans the iron has been shown to be incorporated in hemoglobin when given as a supplement in iron deficiency (Will and Vilter, 1954; Hodgkinson, 1961), and increases in hemoglobin levels have been demonstrated in two fortification studies. Garby and Areekul (1974) used an Fe(III)EDTA-fortified fish sauce in an adult Thai population, and we have demonstrated an improvement in iron status in a controlled trial using Fe(III)EDTA-fortified curry powder in an Indian community (P. MacPhail *et al.*, unpublished data). However, the chelate's peculiar advantages in food fortification over other nonheme iron sources have only recently been recognized (Viteri *et al.*, 1976, 1978; Layrisse and Martinez-Torres, 1977; Martinez-Torres *et al.*, 1979; MacPhail *et al.*, 1981).

A. Absorption of Iron from Fe(III)EDTA

There is a fair amount of information about the absorption of Fe(III)EDTA. As with other forms of nonheme iron, the percentage of the administered dose that is absorbed is influenced by the iron nutritional status of the individual; it has been shown to be directly related to the percentage absorption of ferrous ascorbate and indirectly related to the serum ferritin concentration (Layrisse and Martinez-Torres, 1977; Viteri *et al.*, 1978; Martinez-Torres *et al.*, 1979; MacPhail *et al.*, 1981). This suggests that it is absorbed by the same mechanism as the other forms of nonheme iron, and that it becomes separated from the EDTA chelator in the process. It has, however, been shown that some of the intact chelate is absorbed and, in common with other metal–EDTA complexes, is rapidly excreted in the urine. When the amount of iron absorbed is small, about 14–16% of it is lost in the urine within 24 hours (Martinez-Torres *et al.*, 1979; MacPhail *et al.*, 1981). This is equivalent to less than 1% of the iron ingested.

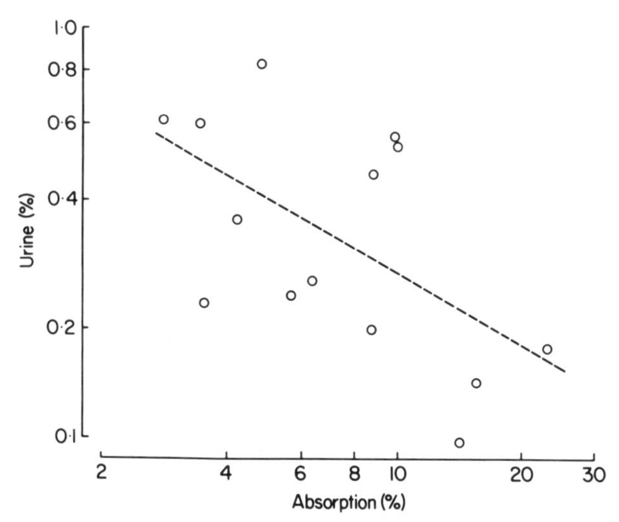

Figure 1. The relationship between the excretion of ^{59}Fe in the urine collected over 24 hours (% dose) and the absorption of ^{59}Fe from $^{59}Fe(III)EDTA$ in aqueous solution ($r = -0.58$, $p < .05$). (From MacPhail *et al.*, 1981, with permission of Cambridge University Press.)

However, an inverse relationship between the amount of iron absorbed and the amount appearing in the urine has been demonstrated, so that less of the iron chelate and more liberated iron is absorbed when absorption is enhanced (MacPhail *et al.*, 1981) (Fig. 1). Studies in humans using [^{14}C]dicalcium EDTA indicate that that chelate is also poorly absorbed from the gastrointestinal tract (less than 5%) and that more than 98% of the absorbed chelate appears unchanged in the urine (Foreman and Trujillo, 1954). Relatively recent work in swine has confirmed that most of the iron is split from the EDTA complex in the lumen of the gut (Candela *et al.*, 1984). Some 5% of this iron was absorbed, but more than 90% was found in the feces in an insoluble form, not associated with the chelate. In these studies about 5% of the ^{14}C derived from [^{14}C]Fe(III)EDTA was absorbed. No correlation was found between the iron and ^{14}C absorbed, which suggested that the iron was not absorbed in the chelated form. However, as in humans, a small amount of iron (less than 1% of the absorbed dose), was detected in the urine, probably in association with EDTA. The remainder of the ^{14}C was eliminated in the feces, largely in a soluble fraction.

The iron is thus largely separated from the EDTA during the absorption process, and yet it is protected to a considerable degree from binding by inhibitory ligands within the intestinal lumen. This can be deduced by comparing the absorption of Fe(III)EDTA with that of $FeSO_4$, in the fasting state and when administered with food. Significantly less of the former is absorbed in the absence of food (Martinez-Torres *et al.*, 1979), but in the presence of cereal or

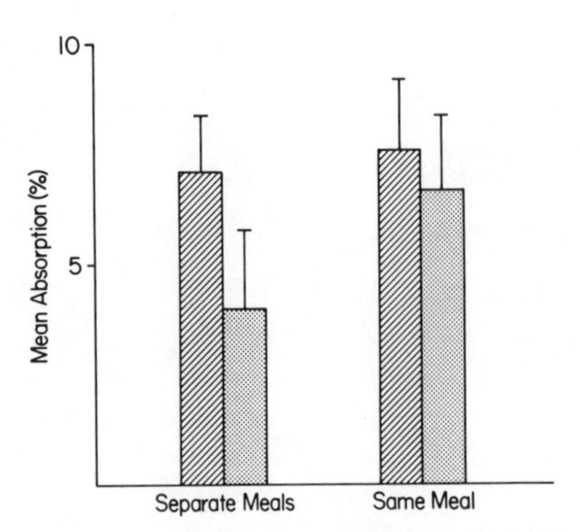

Figure 2. Mean absorption of iron (±1 SD) from Fe(III)EDTA (hatched bars) and $FeSO_4$ (dotted bars) when fed in separate meals and in the same meal. (Data from MacPhail *et al.*, 1981.)

milk, absorption from Fe(III)EDTA remains unchanged while absorption from simple iron salts is depressed (Viteri *et al.*, 1978; Martinez-Torres and Layrisse, 1979; MacPhail *et al.*, 1981). This is illustrated in Fig. 2 where iron absorption from a maize meal fortified with Fe(III)EDTA is shown to be more than twice that from a similar $FeSO_4$-fortified meal. Interestingly enough, the iron in $FeSO_4$ is as well absorbed as that in Fe(III)EDTA when they are fed together in the same meal (Martinez-Torres *et al.*, 1979; MacPhail *et al.*, 1981) (Fig. 2). (The reason for this is discussed later in this section.) These results indicate that while iron absorption from simple iron salts such as $FeSO_4$ is very sensitive to inhibiting substances in food, the absorption from Fe(III)EDTA is not. This effect has been demonstrated using a variety of substances known to inhibit iron absorption. Most of the studies have involved constituents of various standard meals often containing more than one known inhibitor. For example, iron absorption from an Fe(III)EDTA-fortified meal containing desferrioxamine as well as beans, wheat, maize, and coffee was three times greater than when $FeSO_4$ was the fortificant (Viteri *et al.*, 1978). Figure 3 shows the effect on iron absorption of adding known inhibitors to aqueous solutions containing either $FeSO_4$ or Fe(III)EDTA. While bran reduced absorption from $FeSO_4$ 11-fold, it had no significant effect on absorption from Fe(III)EDTA. However, tea, which is known to be a potent inhibitor (Disler *et al.*, 1975), reduced iron absorption from Fe(III)EDTA about sevenfold (MacPhail *et al.*, 1981).

Ascorbic acid has been shown to produce a dose-dependent enhancement of iron absorption from fortified cereals (Derman *et al.*, 1977; Sayers *et al.*, 1973; Björn-Rasmussen and Hallberg, 1974) and also to inhibit the binding of iron to food fiber *in vitro* (Reinhold *et al.*, 1981; Fernandez and Phillips, 1982). A similar enhancement is seen when Fe(III)EDTA and ascorbate are fed in aqueous

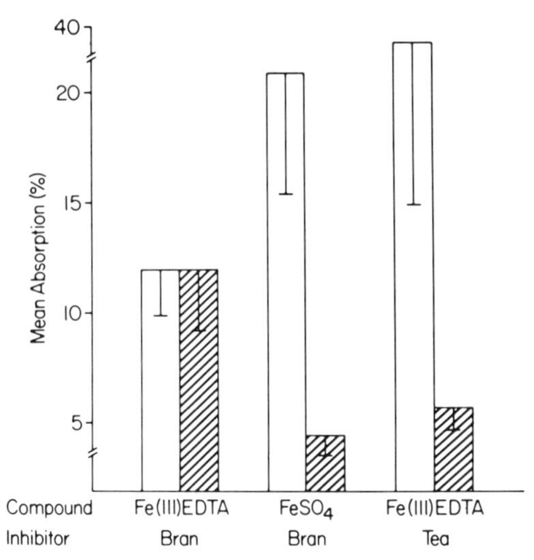

Figure 3. The effect of adding an inhibitor of iron absorption on the absorption of iron from aqueous solutions sweetened with $FeSO_4$- or Fe(III)EDTA-fortified sugar. Values are shown as means (±1 SD). Open bars, without inhibitor; hatched bars, with inhibitor. (Data from MacPhail *et al.*, 1981.)

solutions, although the effect is less marked when an inhibitor in the form of maize is also present. In fact, at higher ascorbate:iron ratios the addition of maize has a greater inhibitory effect than when only Fe(III)EDTA is present (MacPhail *et al.*, 1981). All these observations suggest that competitive binding of iron occurs in the gastrointestinal lumen and that the amount of iron absorbed depends on the quantities and the relative affinities for iron of the various ligands in the meal.

Much of the experimental work on iron fortification and food iron absorption has depended on the observation that the nonheme iron added to a meal and the intrinsic food iron behave to a large extent as a single pool equally susceptible to whatever enhancers and inhibitors of absorption may be present (Cook *et al.*, 1972; Hallberg and Björn-Rasmussen, 1972). The studies using Fe(III)EDTA, however, have revealed a variation on this theme. Although a number of experiments comparing absorption from Fe(III)EDTA with that from either added iron salts or intrinsic food iron show that the ratio of iron absorbed from these sources is close to unity (Layrisse and Martinez-Torres, 1977; Viteri *et al.*, 1978; Martinez-Torres *et al.*, 1979; MacPhail *et al.*, 1981), careful analysis of the data reveals a slight but statistically significant advantage for Fe(III)EDTA (MacPhail *et al.*, 1981). This suggests that the iron in the Fe(III)EDTA complex may not exchange completely with the nonheme iron common pool. However, a considerable degree of exchange does occur: this has been demonstrated by experi-

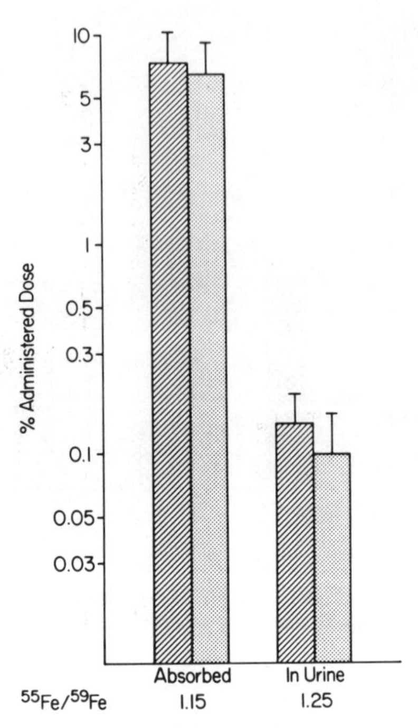

Figure 4. Mean percentages of iron absorbed and appearing in the urine following ingestion of maize porridge fortified with equimolar quantities of ^{55}Fe(III)EDTA (hatched bars) and $^{59}FeSO_4$ (dotted bars). (Data from MacPhail *et al.*, 1981.)

ments in which subjects were given maize porridge fortified with equimolar quantities of $FeSO_4$ and Fe(III)EDTA labeled with different radioisotopes of iron (Fig. 4). The ratio of the two isotopes in the blood, and therefore of the iron absorbed from Fe(III)EDTA and $FeSO_4$, was close to unity (1.15 ± 0.11), and a similar ratio was found in the urine (1.25 ± 0.08). Since the isotope given as $FeSO_4$ could only have appeared in the urine as Fe(III)EDTA, exchange must have taken place. This was confirmed by an *in vitro* experiment in which increased solubilization of iron from $^{59}FeSO_4$-fortified maize porridge was induced by the addition of an equimolar quantity of unlabeled Fe(III)EDTA (MacPhail *et al.*, 1981) (Fig. 5).

B. Fe(III)EDTA and Food Vehicle Fortification

The absorption of iron from a wide variety of vegetable meals fortified with Fe(III)EDTA has been determined. In some instances a consituent of the meals such as sugar or wheat dough was fortified and then used to prepare the meal. Table I shows the mean absorptions from these vehicles, standardized to 40%

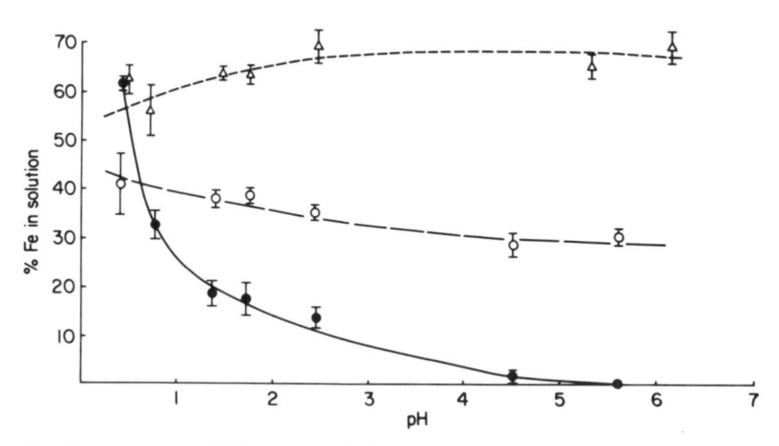

Figure 5. The percentage of ^{59}Fe appearing in the supernatant fraction after incubation of portions of maize porridge fortified with iron at various pH values. (●—● $^{59}FeSO_4$ alone; △- - -△ ^{59}Fe(III)EDTA alone; ○– – ○ equimolar quantities of $^{59}FeSO_4$ and cold Fe(III)EDTA). (From MacPhail *et al.*, 1981, with permission of Cambridge University Press.)

absorption of a ''reference'' dose of ferrous ascorbate to compensate for the differences in the iron nutritional status of the subjects and to bring the values into the range expected in iron deficiency (Hallberg *et al.*, 1978). It can be seen that a daily intake of food fortified with 10 mg of iron as Fe(III)EDTA would increase iron absorption in iron-deficient subjects by between 0.8 and 2.1 mg/day. It has been estimated that increasing absorption by only 1 mg/day would be sufficient to reduce the prevalence of iron deficiency in a female population from 20 to 5% (International Nutritional Anemia Consultative Group, 1977).

One of the major problems in iron fortification has been the development of unacceptable color changes in fortified foods (Sayers *et al.*, 1974). Fe(III)EDTA is a pale yellow water-soluble powder with a high stability constant, and there should thus not be major problems in finding a suitable dietary ingredient to serve as a vehicle for fortification. No important color changes have been reported in the vehicles so far tested. However, adding Fe(III)EDTA-fortified sugar to tea produces the same characteristic blackish discoloration reported with iron salts (Disler *et al.*, 1975). The possibility that changes in Fe(III)EDTA-fortified vehicles stored over prolonged periods may impair consumer acceptability still needs to be investigated.

C. Toxicity of EDTA Chelates

The fact that even small amounts of Fe(III)EDTA are absorbed poses the question whether chronic ingestion of the chelate will produce toxic side effects.

Table I

ABSORPTION OF IRON FROM DIFFERENT VEHICLES FORTIFIED WITH Fe(III)EDTA

Vehicle	Iron absorption			Reference
	Geometric mean (%)	Standardized (%) to 40% reference salt[a]	Projected absorption (mg) at fortification level of 67 mg Fe(III)EDTA (10 mg Fe)/day	
Maize dough	8.5	9.6	0.96	Layrisse and Martinez-Torres (1977)
Maize meal	7.2	8.2	0.82	MacPhail *et al.* (1981)
Wheat	9.6	9.7	0.97	Martinez-Torres *et al.* (1979)
Sweet manioc flour	14.4	16.4	1.64	Martinez-Torres *et al.* (1979)
Milk	11.6	16.7	1.67	Layrisse and Martinez-Torres (1977)
Sugar	5.9	10.7	1.07	Martinez-Torres *et al.* (1979)
Sugar	7.0	8.0	0.80	MacPhail *et al.* (1981)
Sugarcane syrup	5.9	10.7	1.07	Martinez-Torres *et al.* (1979)
Curry powder	12.5	21.3	2.13	MacPhail (1982)

[a]Standardized absorption $= \frac{\text{Fe(III)EDTA absorption}}{\text{reference (Fe ascorbate) absorption}} \times 40$ (Hallberg *et al.*, 1978).

Particular concern has been expressed regarding the possible adverse effects of long-term ingestion of Fe(III)EDTA on calcium balance and on trace metal metabolism. Studies following ingestion of sodium calcium EDTA have failed to show changes in calcium excretion (Spencer, 1960), and since the stability constant of Fe(III) with EDTA (log $K = 25.0$) is considerably greater than Ca (log $K = 10.6$), the likelihood of Fe(III)EDTA chelating calcium in the diet is small. However, the iron probably leaves the EDTA complex during absorption, and the possibility that other metal–EDTA complexes (e.g., zinc) may be formed must be considered; specific attention should be devoted to this when monitoring the consequences of a fortification program. In a controlled double blind fortification trial we were not able to detect any differences in plasma zinc levels between a sample from a group which had received curry powder fortified with Fe(III)EDTA for a period of 3 years and a matched sample from a control group which received unfortified curry powder (P. MacPhail *et al.*, unpublished observations).

While no information regarding the long-term effects of using Fe(III)EDTA as a fortificant is available, sodium calcium EDTA is widely used as a preservative in foods to prevent oxidative damage by free metals. In the United States, EDTA is allowed in certain foods at levels ranging from 25 to 800 mg/kg, and an acceptable daily intake for humans of up to 2.5 mg/kg body weight has been set by the joint Food and Agriculture Organization of the United Nations and World Health Organization Expert Committee on Food Additives (Code of Federal Regulations, 1974; World Health Organization, 1974). Addition of 67 mg of Fe(III)EDTA (10 mg iron) per day to the diet would add only a small increment to the quantity of EDTA already present.

III. HEMOGLOBIN

Meat and meat products have a twofold beneficial effect on iron nutrition. Not only is the iron in the hemoglobin and myoglobin of meat well absorbed, but meat also enhances the absorption of the nonheme iron present in the diet. The individual contribution of meat to daily iron absorption was vividly illustrated by a study reported by Björn-Rasmussen and co-workers (1974), in which they fed a diet containing 16.4 mg nonheme iron and 1 mg heme iron to healthy volunteers (Fig. 6). Despite the small amount of heme iron, it accounted for a third of the total amount absorbed. The mechanisms responsible for the ready availability of heme iron have been intensively studied.

A. Absorption of Heme Iron

During digestion heme is split from the protein, but the iron is not released from the porphyrin ring, the heme molecule being taken up into the intestinal

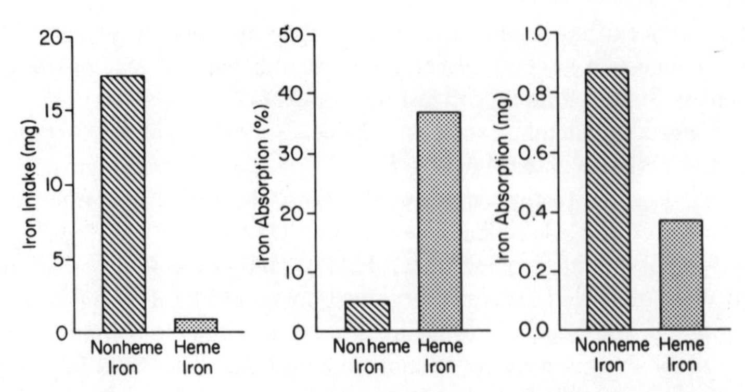

Figure 6. Absorption by 10 subjects of iron from a diet containing 17.4 mg iron/day—16.4 mg nonheme iron and 1 mg heme iron. (Data from Björn-Rasmussen *et al.*, 1974.)

mucosal cells (Conrad *et al.*, 1966, 1967). Because of this, most of the luminal factors that exert such a profound effect on the absorption of the nonheme iron in the diet, and also on the absorption of nearly all iron fortificants, do not affect the absorption of heme. Thus, its absorption is not influenced by the ascorbic acid content of the meal (Turnbull *et al.*, 1962), or the presence or absence of gastric hydrochloric acid (Heinrich *et al.*, 1971), each of which is of critical importance for the absorption of nonheme iron. Whether the heme iron is in the Fe(II) or Fe(III) form is also irrelevant (Heinrich and Gabba, 1977). However, heme iron is much better absorbed when it is ingested as hemoglobin than when the heme is given without the globin (Conrad *et al.*, 1966). In attempting to explain why the presence of globin enhanced the absorption of heme iron, Conrad and co-workers (1966) showed that the absorption was inversely related to the degree of polymerization of the heme. They found that the degradation products of globin prevented this polymerization by binding heme to form monomeric hemochromes and concluded that the enhancement of heme absorption by globin was due to prevention of polymerization within the intestinal lumen.

Hemoglobin iron is thus better absorbed than heme iron administered without the globin, but if the meal contains meat the absorption of hemoglobin iron is further enhanced. Martinez-Torres and Layrisse (1971) found that veal approximately doubled the absorption of hemoglobin iron, whereas Hallberg and co-workers (1979) found that the absorption of heme iron from wheat rolls containing hemoglobin as an iron fortificant was increased by about 60% if a hamburger was eaten with the roll. Martinez-Torres and Layrisse (1971) suggested that the additional protein degradation products from the meat were enhancing the action of the globin degradation products in preventing heme polymerization. However, Hallberg and co-workers (1979) pointed out that meat enhances the absorption of the nonheme iron in a meal to approximately the same extent as it enhances the absorption of the heme iron; they concluded that it was more likely that the effect was a nonspecific one, namely a stimulation of the processes of digestion. This would have to be independent of the effect on the secretion of gastric juice, since Björn-Rasmussen and Hallberg (1979) have shown that the absorption-enhancing effect of meat is seen also in subjects with achylia gastrica. It seems that the nature of the mechanism has not been firmly established. Indeed, there is some debate regarding whether heme iron is absorbed as such when ingested as part of muscle. Some *in vitro* evidence suggests that under such circumstances well-absorbed nonheme iron compounds of low molecular weight are the major degradation products that are released during digestion (Hazell *et al.*, 1978, 1981). It may therefore be necessary to distinguish between the absorption of hemoglobin iron when fed alone and when fed as part of meat. Even with meat there may be variations in bioavailability depending on the mode of preparation. Schricker and Miller (1983) demonstrated a linear rise in the nonheme iron content of meat with increasing cooking times. In addition, reductants such as ascorbic acid, which are used for the preservation of meat, increased the nonheme iron content of cooked meat even further. It was postulated that cooking led to the oxidative degradation of the myoglobin and hemoglobin present in meat. Although the effect was only modest with short cooking times, prolonged cooking at high temperatures led to a 40% decrease in the heme iron content of meat.

After heme has been taken up into the intestinal mucosal cells, the iron is split from the porphyrin ring by the action of heme oxygenase (Raffin *et al.*, 1974). Thereafter it follows the same route as absorbed nonheme iron and is released into the portal circulation. There is evidence that the rate-limiting step in heme iron absorption is the splitting of iron from heme within the mucosal cell (Wheby and Spyker, 1981). The absorption of heme iron, like that of nonheme iron, occurs predominantly in the duodenum and upper jejunum, and heme oxygenase is present in higher concentrations in the cells of these regions. In rats, the heme oxygenase activity increases when the subject is iron deficient, and the rate of heme catabolism in the mucosa is more rapid under these circumstances (Wheby *et al.*, 1970). This is associated with the absorption of increased amounts of heme iron, but subsequent studies in human subjects have made it clear that the iron nutritional status of the individual has much less influence on the percentage absorption of heme iron than it does on the percentage absorption of nonheme iron. This is especially the case with the quantities of heme found in the average diet, since even when the amount of meat eaten is considerable, the daily heme iron consumption is only about 1–2 mg (Hallberg *et al.*, 1979). Hallberg and co-workers (1979) found that at heme iron doses of 0.5, 1.0, and 5.0 mg, blood donors absorbed about 1.5 times as much iron as subjects who were not blood donors but absorbed 3.4 times as much iron when it was in the form of ferrous ascorbate. Heinrich and co-workers (1969) noted that individuals with normal iron stores absorbed 20% of the 0.56 mg iron in 10 μmol of rabbit hemoglobin, and 23% of the same amount of iron given as ferrous ascorbate; in borderline iron

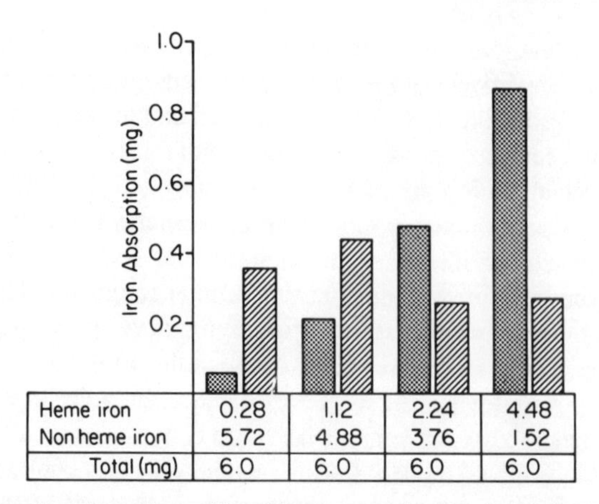

Figure 7. Mean absorption of heme (dotted bars) and nonheme iron (hatched bars) from a meal of gravy and potato in which the total iron content was kept constant but the relative amounts of heme and nonheme iron were varied. (Data from Bezwoda *et al.*, 1983.)

deficiency these figures were 30 and 81%, respectively. In one study we varied the amounts of heme and nonheme iron in a meal of mashed potato and gravy while keeping the total iron content constant, and found no relationship between iron nutritional status and heme iron absorption (Fig. 7) (Bezwoda *et al.*, 1983). In fact, it was noteworthy that there was very little variation in the percentage absorption of the heme iron at the different heme:nonheme iron ratios: the mean values ranged from 19.0 to 21.4% as the heme iron content rose from 0.28 to 4.48 mg, so that the amount of iron absorbed increased 16-fold. In contrast, the nonheme iron absorption varied widely, and the mean was 18% when the meal contained 1.52 mg nonheme iron and 6% when the quantity was increased to 5.72 mg; the amount of nonheme iron absorbed thus hardly changed. Hallberg and co-workers (1979) also observed no fall in the percentage absorption of heme iron over a dosage range from 0.5 to 5.0 mg, that is, a 10-fold rise in the amount of iron absorbed as the amount of hemoglobin was increased 10-fold. The relationship between the amount of heme iron ingested and the amount absorbed does not continue indefinitely, however, since a mean of only 3.8% of the 43 mg heme iron in a blood sausage meal was absorbed. Interestingly, at this very high dose there was a good correlation between the absorptions of heme iron and of 3 mg iron as ferrous ascorbate in the fasting state; the latter has been widely used as a measure of the individual's iron nutritional status, and correlates well with the plasma ferritin level and other indices of the body iron reserve.

B. Hemoglobin Iron as a Fortificant

In considering hemoglobin as a possible iron fortificant, two properties offer very considerable advantages: (1) its absorption appears to vary remarkably little from subject to subject whatever the vegetable composition of the meal or the individual's ability to secrete hydrochloric acid in the gastric juice, and (2) the predictable percentage of the added iron that is absorbed. Thus, it seems probable that the daily iron absorption could be increased by about 1 mg if the average diet could be fortified with 5 mg of hemoglobin iron. Account would, however, have to be taken of the other feature of heme iron absorption, namely that in contrast to nonheme iron absorption, there is relatively little difference between the amounts absorbed by iron-deficient and iron-replete subjects. Any adjustment in total daily iron absorption by the latter would therefore have to be largely achieved by reducing the amount of nonheme iron absorbed. The margin for adjustment would be further reduced in those individuals whose meat intake was high, since they would absorb some of the heme in the meat as well as a greater percentage of the fortification heme. Such subjects would be at real risk of the gradual accumulation of tissue iron stores that might eventually reach toxic levels. The ideal solution to the problem of avoiding iron overload in the nutritionally privileged members of the community while effectively countering iron deficiency in the vulnerable members would, of course, be to fortify some dietary ingredient consumed only by the latter. This would be more easily achieved in the case of the one vulnerable group (infants and children) than the other (the women from puberty to menopause).

Reizenstein (1975) has pointed out that only one-tenth of all the hemoglobin produced by Swedish slaughterhouses is used for food production. He has drawn attention to the waste of protein as well as of iron, and has reported that hemoglobin can be used to fortify bread and sausage so as to produce a product that appears to be satisfactory from the point of view of taste, consistency, and appearance. The sausage contained 15 mg of heme iron/100 g. In Chile, dried bovine hemoglobin is being used as an iron fortificant (A. Stekel, personal communication). It has been added to a rice-based cereal formula that is used as a supplementary feed for breast-fed infants. It has also been added to cookies that are consumed by preschool and schoolchildren as part of a school lunch program. Thirty grams of cookies contain about 5 mg of hemoglobin iron. Radioisotopic studies have confirmed that about 20% of this iron is absorbed. By confining the fortification to the preadult members of the population, the theoretical danger of iron overload has been avoided.

There thus appear to be several possible vehicles that could be fortified with hemoglobin iron. Because of its color, there would appear to be some restriction of the range of possible vehicles, and shelf life may also prove to be a problem, although the evidence to date obtained in Chile suggests that slaughterhouse blood that has been spray-dried shortly after collection is stable and satisfactory. However, it must be accepted that a good deal of technical information on the standards required for the sterile collection, drying, and storage of animal blood is still needed. There may also be other practical restrictions on the use of such material, since the quantity of hemoglobin available must in itself be a function of the per capita meat consumption, and this option may not be open to many developing countries where little meat is consumed. The alternative of exporting dried material from developed countries may prove prohibitively expensive.

IV. CONCLUSIONS AND SUMMARY

Both Fe(III)EDTA and hemoglobin may have potential roles to play as iron fortificants. Their major advantage lies in the fact that they are well absorbed from diets of low iron availability. The iron in Fe(III)EDTA is less susceptible to the inhibitory ligands present in the predominantly cereal diets consumed by a large proportion of the world's population, and there is some evidence that the overall absorption from such diets is increased when the chelate is present. The advantage of hemoglobin as an iron fortificant lies in the fact that it is absorbed into mucosal cells as heme and not as iron, and is thus unaffected by the many ligands in the diet that inhibit iron absorption. The fact that the percentage absorption of hemoglobin iron in physiologic doses is relatively unaffected by the iron status of the individual does, however, suggest the need for caution in its application as a fortificant, since iron-replete individuals might be expected to absorb just as much as those who are iron deficient. This means that its potential applications should be confined to specific vulnerable target groups.

From a practical standpoint, Fe(III)EDTA can be added unobtrusively to several food vehicles without causing changes in color or taste. However, long-term field trials using several different vehicles fortified with Fe(III)EDTA are still needed to confirm its usefulness and to ensure its safety. In this latter regard, special attention should be directed to its possible effect on trace mineral metabolism.

Relatively little information is yet available on the degree to which hemoglobin can meet the exacting requirements of a safe iron fortificant. The problem is compounded by the fact that it is a biological material, and therefore other considerations arise, such as those of sterility. It nevertheless deserves further study, especially since this valuable source of both iron and protein is going to waste in vast amounts in slaughterhouses throughout the world.

ACKNOWLEDGMENT

The authors are grateful for support from the South African Atomic Energy Board.

REFERENCES

Bezwoda, W. R., Bothwell, T. H., Charlton, R. W., Torrance, J. D., MacPhail, A. P., Derman, D. P., and Mayet, F. (1983). The relative dietary importance of haem and non-haem iron. *S. Afr. Med. J.* **64,** 552–556.

Björn-Rasmussen, E., and Hallberg, L. (1974). Iron absorption from maize. Effect of acid on iron absorption from maize supplemented with ferrous sulphate. *Nutr. Metab.* **16,** 94–100.

Björn-Rasmussen, E., and Hallberg, L. (1979). Effect of animal proteins on the absorption of food iron in man. *Nutr. Metab.* **23,** 192–202.

Björn-Rasmussen, E., Hallberg, L., Isaksson, B., and Arridsson, B. (1974). Food iron absorption in man. Applications of the two-pool extrinsic tag method to measure haem and non-haem iron absorption from the whole diet. *J. Clin. Invest.* **53,** 247–255.

Bothwell, T. H., Charlton, R. W., Cook, J. D., and Finch, C. A. (1979). "Iron Metabolism in Man." Blackwell, Oxford.

Candela, E., Camacho, M. K., Martinez-Torres, C., Perdoma, J., Mazzarri, G., Acurero, G., and Layrisse, M. (1984). Iron absorption by humans and swine from Fe(III)EDTA. Further studies. *J. Nutr.* (In press).

Code of Federal Regulations. (1974). Title 21, Part 121. U.S. Govt. Printing Office, Washington, D.C.

Conrad, M. E., Weintraub, L. R., Sears, D. A., and Crosby, W. H. (1966). Absorption of hemoglobin iron. *Am. J. Physiol.* **211,** 1123–1130.

Conrad, M. E., Benjamin, B. I., Williams, H. L., and Foy, A. L. (1967). Human absorption of hemoglobin iron. *Gastroenterology* **53,** 5–10.

Cook, J. D. (1977). Absorption of food iron. *Fed. Proc. Fed. Am. Soc. Exp. Biol.* **36,** 2028–2032.

Cook, J. D., Layrisse, M., Martinez-Torres, C., Walker, R. B., Monsen, E., and Finch, C. A. (1972). Food iron absorption measured by an extrinsic tag. *J. Clin. Invest.* **51,** 805–815.

Derman, D., Sayers, M., Lynch, S. R., Charlton, R. W., Bothwell, T. H., and Mayet, F. (1977). Iron absorption from a cereal diet containing cane sugar fortified with ascorbic acid. *Br. J. Nutr.* **38,** 261–269.

Disler, P. B., Lynch, S. R., Charlton, R. W., Torrance, J. D., Bothwell, T. H., Walker, R. B., and Mayet, F. (1975). The effect of tea on iron absorption. *Gut* **16,** 193–200.

Fernandez, R., and Phillips, S. F. (1982). Components of fiber bind iron *in vitro*. *Am. J. Clin. Nutr.* **35,** 100–106.

Foreman, H., and Trujillo, T. T. (1954). The metabolism of C^{14} labelled ethylenediaminetetraacetic acid in human beings. *J. Lab. Clin. Med.* **43,** 566–571.

Garby, L., and Areekul, S. (1974). Iron supplementation in Thai fish sauce. *Ann. Trop. Med. Parasitol.* **68,** 467–476.

Gillooly, M., Bothwell, T. H., Torrance, J. D., MacPhail, A. P., Derman, D. P., Bezwoda, W. R., Mills, W., and Charlton, R. W. (1983). The effects of organic acids, phytates and polyphenols on the absorption of iron from vegetables. *Br. J. Nutr.* **49,** 331–342.

Hallberg, L., and Björn-Rasmussen, E. (1972). Determination of iron absorption from whole diet. A new two-pool model using two radioiron isotopes given as haem and non-haem iron. *Scand. J. Haematol.* **9,** 193–197.

Hallberg, L., Björn-Rasmussen, E., Garby, L., Pleehachinda, R., and Suwanik, R. (1978). Iron absorption from South-East Asian diets and the effect of iron fortification. *Am. J. Clin. Nutr.* **31,** 1403–1408.

Hallberg, L., Björn-Rasmussen, E., Howard, L., and Rossander, L. (1979). Dietary heme iron absorption. A discussion of possible mechanisms for the absorption-promoting effect of meat and for the regulation of iron absorption. *Scand. J. Gastroenterol.* **14,** 769–779.

Hazell, T., Ledward, D. A., and Neale, R. J. (1978). Iron availability from meat. *Br. J. Nutr.* **39,** 631–638.

Hazell, T., Ledward, D. A., Neale, R. J., and Root, I. C. (1981). The role of non-heme proteins in meat hemoprotein digestion. *Meat Sci.* **5,** 397–405.

Heinrich, H. C., and Gabbe, E. E. (1977). Hemoglobin-iron for the prophylaxis and treatment of iron deficiency. *Klin. Wochenschr.* **55,** 1043–1049.

Heinrich, H. C., Bartels, H., Gabbe, E. E., Meineke, B., Nass, W. P., and Whand, D. H. (1969).

Die intestinale resorption des Nahrungs-eisens aus dem Hämoglobin, der Leber und Muskulatur bei Menschen mit normalen Eisenreserven and Personen mit prälatentem/latentem Eisenmangel. *Klin. Wochenschr.* **47,** 309–317.

Heinrich, H. C., Gabbe, E. E., and Kugler, G. (1971). Nahrungs-Eisen resorption aus Schweine-Fleisch, -Leber und Hämoglobin bei Menschen mit normalen und erschöpften Eisen reserven. *Klin. Wochenschr.* **49,** 819–825.

Hodgkinson, R. (1961). A comparative study of iron absorption and utilization following ferrous sulphate and sodium ironedetate (''Sytron''). *Med. J. Aust.* **48,** 809–811.

International Nutritional Anemia Consultative Group (INACG). (1977). ''Guidelines for the Eradication of Nutritional Anemia,'' Nutrition Foundation, New York.

Layrisse, M., and Martinez-Torres, C. (1977). Fe(III)EDTA complex as iron fortification. *Am. J. Clin. Nutr.* **30,** 1166–1174.

MacPhail, A. P. (1982). An approach to the evaluation of iron status and the prevention of iron deficiency. Ph.D. Thesis, University of the Witwatersrand, Johannesburg.

MacPhail, A. P., Bothwell, T. H., Torrance, J. D., Derman, D. P., Bezwoda, W. R., Charlton, R. W., and Mayet, F. (1981). Factors affecting the absorption of iron from Fe(III)EDTA. *Br. J. Nutr.* **45,** 215–227.

Martinez-Torres, C., and Layrisse, M. (1971). Iron absorption from veal muscle. *Am. J. Clin. Nutr.* **24,** 531–540.

Martinez-Torres, C., Romano, E. L., Renzi, M., and Layrisse, M. (1979). Fe(III)EDTA complex as iron fortification. Further studies. *Am. J. Clin. Nutr.* **32,** 809–816.

Narasinga Rao, B. S., and Vijayasarathy, C. (1975). Fortification of common salt with iron. Effect of chemical additives on stability and bioavailability. *Am. J. Clin. Nutr.* **20,** 1395–1401.

Raffin, S. B., Woo, C. H., Roost, K. T., Price, D. C., and Schmid, R. (1974). Intestinal absorption of hemoglobin; iron heme cleavage by mucosal heme oxygenase. *J. Clin. Invest.* **54,** 1344–1352.

Reinhold, J. G., Garcia, J. S., and Garzon, P. (1981). Binding of iron by fiber of wheat and maize. *Am. J. Clin. Nutr.* **34,** 1384–1391.

Reizenstein, P. (1975). Cattle haemoglobin—a possible dietary iron supplement. *Br. J. Haematol.* **31,** 265–268.

Sayers, M. H., Lynch, S. R., Jacobs, P., Charlton, R. W., Bothwell, T. H., Walker, R. B., and Mayet, F. (1973). The effects of ascorbic acid supplementation on the absorption of iron in maize, wheat and soya. *Br. J. Haematol.* **24,** 209–218.

Sayers, M. H., Lynch, S. R., Charlton, R. W., Bothwell, T. H., Walker, R. B., and Mayet, F. (1974). The fortification of common salt with ascorbic acid and iron. *Br. J. Haematol.* **28,** 483–495.

Schricker, B. R., and Miller, D. D. (1983). Effects of cooking and chemical treatment on heme and nonheme iron in meat. *J. Food. Sci.* **48,** 1340–1350.

Spencer, H. (1960). Studies of the effect of chelating agents in man. *Ann. N.Y. Acad. Sci.* **88,** 435–449.

Stewart, I., and Leonard, C. D. (1952). Chelates as sources of iron for plants growing in the field. *Science* **116,** 564–566.

Turnbull, A. L., Cleton, F., and Finch, C. A. (1962). Iron absorption. IV. The absorption of hemoglobin iron. *J. Clin. Invest.* **41,** 1898–1907.

Viteri, F. E., Torun, B., and Garcia, R. (1976). Iron fortification of sugar: Advantages of NaFe EDTA over $FeSO_4$. *Abstr. Symp. Free Commun., Int. Congr. Nutr., 10th, 1975,* No. 8211.

Viteri, F. E., Garcia-Ibañez, R., and Torun, B. (1978). Sodium iron NaFe EDTA as an iron fortification compound in Central America. Absorption studies. *Am. J. Clin. Nutr.* **31,** 961–971.

Wheby, M. S., and Spyker, D. A. (1981). Hemoglobin iron absorption kinetics in the iron-deficient dog. *Am. J. Clin. Nutr.* **34,** 1686–1693.

Wheby, M. S., Suttle, G. E., and Ford, K. T. (1970). Intestinal absorption of hemoglobin iron. *Gastroenterology* **58,** 647–654.

Will, J. J., and Vilter, R. W. (1954). A study of the absorption and utilization of an iron chelate in iron deficient patients. *J. Lab. Clin. Med.* **44,** 499–505.

World Health Organization. (1974). Toxicological evaluation of certain food additives with a review of general principles and of specifications. *W.H.O. Tech. Rep. Ser.* **539.**

III

PRODUCT APPLICATION

6

Wheat and Blended Cereal Foods

FRED BARRETT

Nutrition and Agribusiness Group
USDA, Office of International Cooperation and Development
Washington, D.C.

PETER RANUM

Pennwalt Corporation
Flour Service Department
Buffalo, New York

I. CEREAL FORTIFICATION

Fortification is the process whereby nutrients are added to foods to maintain or improve the quality of the diet of a group, a community, or a population (World

ISBN 0-12-177060-5

Health Organization, 1971). Its purpose may be simply to increase the nutritional value of a particular product, or it may be the basis for a general nutritional intervention program. In this chapter fortification is viewed in its broadest context, that is, a public health measure set up and regulated by the government to improve and maintain the nutritional health of a target population through the planned addition of deficit nutrients to a food staple.

The foods most often considered for use in fortification programs are the basic cereal staples: wheat, rice, corn, and sorghum. There are a number of reasons for this. Nearly everyone in the world consumes foods made from cereal grains. One-half of the calories and protein consumed by the people of the world comes from cereals (Food and Agriculture Organization, 1970–1980). The nutritional significance of cereal grains is even greater in developing countries, where the vast majority of the population depends on cereals for its daily sustenance.

Cereals, especially milled wheat flour and cornmeal, make ideal vehicles for use in fortification programs. Nearly any nutrient can be added—protein or amino acids, vitamins, and minerals—without affecting product quality or acceptability.

Milled cereals allow for the economical and efficient addition of nutrients, because they are generally centrally processed in large, often continuous milling operations. The technology for fortifying cereals is well established. Mills in England, Canada, and the United States have been routinely fortifying flour since the early 1940s. However, this does not mean that a centrally processed commodity is essential. Studies in Guatemala and Thailand have shown that village-level fortification can be operationally feasible (Austin, 1979).

Cereals make especially good vehicles for supplying deficient nutrients like iron in that they are consumed by all classes of people, especially the lower economic classes where the nutritional need is the greatest. They provide a relatively even meal distribution for the added nutrients, since they are generally consumed throughout the day. This ensures an even supply of the nutrients to each individual rather than a single high dose.

One justification for fortifying milled cereals is to replace those nutrients "lost" in the milling process and for which there is evidence of potential risk of deficiency within a population. This type of fortification is often referred to as enrichment.

II. WHEAT CONSUMPTION

Wheat supplies 20% of the world's caloric and protein intake (Food and Agriculture Organization, 1970–1980). The amount of wheat consumed in different countries varies widely, as shown in Table I, from practically none in those countries where it is not available for economic or agronomic reasons to very high, as in Turkey where it supplies over 80% of the calories.

Table I
WHEAT CONSUMPTION IN VARIOUS COUNTRIES OF THE WORLD[a]

Country	Per capita consumption (kg/person/year)	Country	Per capita consumption (kg/person/year)
Afghanistan	119	Mexico	23
Argentina	87	Netherlands	70
Austria	78	Peru	36
Canada	59	Philippines	11
Chile	110	Poland	73
Colombia	10	Rumania	124
Costa Rica	30	Spain	104
Egypt	87	Sudan	8
Ecuador	22	Syria	114
France	95	Turkey	188
Iran	117	United States	54
Ireland	102	Uganda	3
Israel	110	West Germany	54
Japan	26	Yugoslavia	152

[a]Taken from Tweet (1979).

III. IRON CONTENT OF WHEAT

The iron content of wheat itself has been reported to range from 20 up to 100 ppm (Ackroyd and Doughty, 1970; Oparin and Oparina, 1969; Murphy and Law, 1974; Yadav *et al.*, 1973; Ziegler and Greer, 1971). Several studies (Lorenz and Loewe, 1977; Pomeranz and Dikeman, 1982) report levels, on a 14% moisture basis, averaging around 35 ppm.

There is a strong environmnetal influence on the trace mineral content of cereal grains and some suggestion of a genetic influence (Ghanbari and Mameesh, 1971; Bassiri and Nahapetian, 1977; El Gindy *et al.*, 1957; Kleese *et al.*, 1968; Koivistoinen *et al.*, 1974; Nahapetian and Bassiri, 1976; Rasmusson *et al.*, 1971). Hard-wheat varieties contain more iron than soft wheats (Lorenz and Loewe, 1977). In general, the higher the protein in the wheat, the higher its iron content. Work is currently in progress to breed wheats having higher mineral contents, but it remains to be seen whether such efforts will prove to be nutritionally meaningful.

IV. DIETARY IRON CONTRIBUTION OF WHEAT

Wheat contains sufficient iron for it to make, theoretically at least, a highly significant contribution of iron to the diet. The Index of Nutritional Quality of Hansen (Wyse *et al.*, 1976), the ratio of the percentage of nutrient requirement to

Table II
IRON CONTRIBUTION OF WHEAT AND WHEAT FLOUR VERSUS PER CAPITA WHEAT CONSUMPTION

	Per capita consumption (kg/person/year)		
Iron contribution	50	100	150
Whole-wheat or enriched flour (36.5 ppm Fe)			
Amount of Fe/day (mg)	5	10	15
Percentage of female iron requirement (18 mg/day)	28	56	83
Percentage of male iron requirement (10 mg/day)	50	100	150
Unenriched wheat flour (11 ppm Fe)			
Amount of Fe/day (mg)	1.5	3.0	4.5
Percentage of female requirement	8	16	24
Percentage of male requirement	15	30	45

the percentage of energy requirement, for iron in wheat is around 1.0. This means that if the caloric requirements were met through the consumption of wheat, the iron requirements would be met as well. Table II shows the percentage of the dietary iron requirement for males (10 mg/day) and females (18 mg/day) supplied by three different wheat consumption levels. This should be compared to actual wheat consumption levels, as given in Table I.

There were 479 million metric tons of wheat grown throughout the world during 1983 (Anonymous, 1983). At an average iron content of 35 ppm iron, this much wheat would supply 16,730 tons of iron, or 10 mg iron per person per day, to the 4.5 billion world population. The actual contribution of wheat toward meeting dietary iron requirements is of course much lower because of the reduction in iron content that occurs with milling refinement. There is also a question on the bioavailability of indigenous iron in wheat. Whole wheat contains high levels of fiber and phytates that can adversely affect mineral availability when it is the dominant source of food in the diet (Björn-Rasmussen, 1974; Callender and Warner, 1979; Dobbs and Baird, 1977; Edwards *et al.*, 1971; Elwood *et al.*, 1979; McCance and Widdowson, 1942; Miller, 1977).

V. IRON FORTIFICATION OF WHOLE-GRAIN WHEAT

The fortification of wheat is normally done either at the mill or at the bakery. Although it is not currently done, it is technically feasible to fortify whole-grain wheat, and there may be circumstances where such fortification could be useful.

One example is a developed wheat-exporting country providing wheat to a developing country. The addition of iron to the wheat at a single terminal elevator in the exporting country may be cheaper and easier to control, finance, and administer than adding iron at mills within the receiving country. The technology involved would be similar to that used in the fortification of any whole-grain cereal, like rice and bulgur. The structure and appearance of the grain has to be preserved, so that the wheat could be handled and processed in a normal manner. The added iron must have acceptable bioavailability. It must not segregate out or be removed from the wheat during conveying and cleaning. Finally, the iron must end up in the flour, rather than the feed, if it is to do any good.

There have been a number of ways suggested for fortifying whole-grain cereals (Graham *et al.*, 1968); some of these may be applicable to adding iron to wheat. In most of these methods a few highly fortified kernels are produced and added to the grain at rates ranging from 5 to 0.05%. Methods for making a highly fortified kernel include the following:

1. Coating. An iron compound is sprayed on the wheat. This may involve a binder to keep it attached and a protectant to prevent it from being rubbed off during handling.
2. Infusion. The wheat is soaked in an iron solution and dried.
3. Extruded grain analogs. The iron is trapped in a starch or protein matrix using an extrusion process. The final product should resemble a wheat kernel in size, shape, and density.

Increases in the iron content of wheat, whether achieved through selective breeding or fortification, would be of value only if there is no detrimental effect on milling and baking properties and if the additional iron in the wheat is transferred to the flour on milling. A few preliminary studies have shown wheat fortification to be technically possible. How useful it would be as a nutritional intervention program remains to be determined.

VI. MILLING

Before wheat is consumed it is first milled into flour. The milling process involves cleaning the wheat, grinding it into flour, and removing from the flour those fractions of the wheat kernel that are detrimental in a particular end use for which the flour is intended.

A multitude of bakery products are made out of wheat flour. Not only are there yeast-leavened breads but also crackers, cookies, cakes, and other breadstuffs that require varied degrees of refinement of the flour in order for acceptable products to be produced.

Unfortunately, most of the nutrients in wheat are concentrated in the outer layers of the kernel. These are the same portions most detrimental to many

Table III
IRON CONTENT OF FLOUR MILL PRODUCTS[a]

Milling stream		
Type of product	Percentage of total	Iron content (ppm)
Flour (first Mids)	10	8
Flour (second siz + second, third, and fourth Mid)	31	9
Flour (first siz + fifth Mids + third Bk + Bk dust)	16	12
Flour (first, second, and fourth Bk + sixth Mid)	10	16
First clear flour (fifth Bk + tailings)	5	30
Second clear flour	4	40
Tail shorts (red dog)	—	171
Head shorts (fine bran)	7	209
Bran	7	198

[a]From C. A. Watson *et al.* (unpublished data).

baking processes, and they are removed for that reason. This causes the nutrient content of the flour to be lower than that of wheat; the degree to which it is lowered depends on the extraction rate.

Milling does not result in an actual "loss" of nutrients, although that is the way it is usually stated. Rather, it results in a separation of nutrients, removing much of them from direct human consumption but not from human benefit. Those fractions of the wheat kernel removed from flour are generally used as animal feed. A growing percentage of this fraction, which contains higher levels of nutrients than contained in wheat, is finding use in such products as breakfast cereals and other foods where further processing allows it to be used without being detrimental to product quality or consumer acceptability. The extent to which milling separates iron into various mill fractions is illustrated in Table III. Nine different mill streams were collected on a pilot mill. A straight-grade flour, consisting of the first six streams, contained 13 ppm iron. This flour would have an extraction rate of 76%. Further refinement of the flour could be achieved by taking only the first five streams to make a 95% bakers' patent. This flour would have 12 ppm of iron. The first four streams gives an 88% patent flour containing 12 ppm of iron. The final three streams of shorts and bran contain, in this case, 82% of the iron. They would not generally be included in flour, except for whole-wheat flour in which all the streams are combined. In countries using extraction rates higher than 75%, some of the shorts do become incorporated into the flour. It is possible to regrind the shorts and add them back to the flour without causing great problems in bread-baking quality.

VII. IRON CONTENT OF WHEAT FLOUR AND BREAD

The iron content of wheat flour is primarily determined by the extraction rate used in milling the flour; the higher the extraction rate the higher the iron content. In one study of 65 commercially milled wheat flours from North America (Lorenz *et al.*, 1980), the mean iron content was 11.2 ppm with a standard deviation of ±3.8 ppm. The extraction rate used averaged 76% and never exceeded 80%. In another survey of 95 commercially milled flour samples collected from 30 different countries (Ranum *et al.*, 1980), two-thirds of the samples were reported to be made with extraction rates near 76%. Their mean iron content was 13.6 ppm. Extraction rates over 80% were reported for 20% of the flour samples. These had a higher iron content, averaging 30 ppm.

The iron content of the whole-wheat kernel does not have a great effect on the iron content of the flour. This is because the minerals in wheat are primarily associated with the bran. Small wheat kernels, caused by environmental stress or genetic factors, will have a greater proportion of bran to endosperm than larger kernels. As a result the wheat will have a higher iron content, but the iron content of the flour will be normal.

There is evidence of environmental and varietal factors influencing the iron content of flour. A study by Peterson *et al.* (1983) showed a high positive correlation between the flour protein and iron level.

Most of the iron in bread comes from the flour; the rest comes from the yeast. Compressed yeast contains 48 ppm iron. When used in a typical bread at a 2% level it will contribute around 0.6 ppm iron to the bread. Unenriched flour will contribute around 7.6 ppm iron. Other baking ingredients that contain iron include milk or milk substitutes made from whey and soy, but their iron contribution is very small.

Unenriched wheat flour, although it contains about one-third of the iron content of whole wheat, still makes an important contribution to the dietary iron intake in countries with significant wheat consumption, as shown in Table II. This contribution is not enough to fulfill the dietary iron requirement; however, it could be fulfilled if the flour was fortified with iron.

VIII. IRON FORTIFICATION OF WHEAT FLOUR AND BREAD

A. Iron Source

One of the major problems in iron fortification is the choice of the form of iron to add. Factors affecting the choice of the forms of iron to use include bioavailability, functional and stability properties, commercial availability of food grade materials, and cost. Unfortunately, the forms that show the greatest functional stability are often the least assimilable, while those forms that show the greatest bioavailability have the greatest potential for damaging product quality.

1. BIOAVAILABILITY

Iron sources differ in how well they are utilized by the body. There is much controversy on this point, but some general conclusions can be reached on the relative merits of the commonly used iron sources. Ferrous sulfate has good bioavailability. It is the iron source of choice and should be used when functional and stability considerations allow it. The insoluble ferric salts, like ferric orthophosphate and sodium iron pyrophosphate, generally show poor bioavailability. Their use should be discouraged for iron fortification of flour, since better sources are available. The bioavailability of reduced iron powders varies greatly. It is somewhere between that of ferrous sulfate and the insoluble ferric salts. Since it has good functional and stability properties, it is frequently used. Reduced iron is a good compromise between bioavailability and functional considerations, at least until a better source can be found. Other iron sources whose bioavailability is near that of ferrous sulfate, like ferrous fumarate, ferrous gluconate, and ferric ammonium citrate, are more costly to use and not considered economically practical for general fortification of flour.

2. INVISIBILITY

The ideal iron compound should cause no change in color, taste, preparation methods, or appearance in the fortified food. Its presence should be invisible if it is to be used successfully in a general fortification program.

Ferrous sulfate is considered a fairly reactive compound that can hasten the development of oxidative reactions resulting in detrimental off-flavors, colors, or odors. It can react with phenolic compounds to form blue–black colors and darken food products containing large amounts of tannin, like chocolate and barley flour (Waddel, 1973).

Ferrous sulfate, when used in enriched bakers' salt or added to bakery flour, has been known to cause small black spots on the crusts of rolls and bread made by short fermentation methods (Anonymous, 1972). These spots appear to be partially undissolved coarse particles of ferrous sulfate that have undergone surface oxidation. The spot usually disappeared with time as water migrated into the crust. This problem does not occur if the ferrous sulfate used has a fine particle size (85% minimum through a 325 U.S. standard mesh), giving it a faster rate of solution. This type of ferrous sulfate is often called "bakery grade."

There were reports that the addition of high iron levels (88 ppm) to flour caused problems with bread quality and loaf volume when ferrous sulfate was used. This apparently was due to reactions between potassium bromate, a commonly used dough conditioner, and ferrous sulfate resulting in the underoxidation of the dough (Clauss and Kulp, 1974; Jaska and Redfern, 1975). This effect could be prevented by rescheduling the addition of these two ingredients. The problem has not been observed at normal iron addition rates of 25–35 ppm.

There have been a number of reports that ferrous sulfate can promote the development of rancidity in flour, thus aversely affecting the smell and taste of breadstuffs made with such flour (Anonymous, 1975).

In one study (Harrison *et al.*, 1976), samples of unbleached flour enriched with different iron sources were stored at 50 and 23°C. At 50°C a rancid smell was detected with ferrous sulfate after 4 days, and after 11–28 days with reduced iron. All samples stored at 23°C were acceptable after 24 months. In another study (Anonymous, 1968), bread and cake flours enriched with different iron sources, including ferrous sulfate, were stored at room temperature and at 27°C. Baking tests and peroxide values, indicative of oxidative rancidity, run after 3 and 6 months of storage showed no differences that could be attributed to a single iron source. A Norwegian report (Schulerud, 1974) stated that a master mix (3% iron in flour) developed a rancid smell within a few hours with ferrous sulfate and that ferrous sulfate added at 3 or 9 mg Fe/100 g to flour caused rancidity and taste deterioration. However, the report states that a large-scale trial by nine Norwegian flour mills adding ferrous sulfate at 9 mg Fe/lb, to all flour for 19 months produced no complaints from bakers. Other studies report reduced iron to be better than ferrous sulfate in not causing off-odors or taste problems (Fritz *et al.*, 1975; Martin and Halton, 1964; Widh, 1970). These studies show that under certain conditions (high storage temperatures and high iron levels) ferrous sulfate can cause oxidative rancidity problems in flour, but that under normal practice the problem is not severe enough to be noticeable. Ferrous sulfate is routinely used in the United States to enrich bakery flour and flour used in pasta.

Following are some general conclusions regarding the uses of ferrous sulfate:

1. Ferrous sulfate (dried, bakery grade) is the iron source of choice when added at the bakery.
2. Ferrous sulfate can also be used in bakery flour provided that the iron levels added are not too high (above 40 ppm) and that the flour is stored no more than 3 months at moderate temperature (less than 30°C) and humidity conditions.
3. Ferrous sulfate is not recommended for flour that must be stored for extended periods (such as family or all-purpose flour intended for home use) or for flour used in mixes containing fats, oils, and other ingredients that might undergo oxidative rancidity.
4. Ferrous sulfate should not be mixed with wheat flour to form a concentrated premix or master blend for later addition to flour or dough.

Some work has been done on ferrous sulfate to minimize its deleterious oxidative effects while retaining its high bioavailability. One method involves the preparation of finely divided crystals of ferrous sulfate monohydrate coated with ferrous sulfate heptahydrate (Bell, 1974). This product, called "stabilized" ferrous sulfate or Bio-Iron®, showed somewhat improved keeping qualities when added to flour. It is no longer commercially available. Another method is to encapsulate the ferrous sulfate in a hard fat (Jackel, 1971) or other type of coating to prevent it from reacting.

One way of reducing the chemical reactivity of iron is by attaching it to a coordinating agent, such as EDTA, sodium hexametaphosphate, phosphoric acid, or sodium acid sulfate. These agents may also improve its bioavailability.

The physical color of the iron compounds themselves may carry over into the fortified product. Reduced iron powders are dark gray in color. These powders do cause a slight graying effect on the color of flour and bread crumbs, but this is not considered to be a serious problem. The color of ferrous sulfate ranges from off-white to white; it does not change the color of flour. When water is added, ferrous sulfate oxidizes to one or more ferric compounds whose color ranges from gray to green and yellow. This color is not visually apparent in the dough or bread crumb, but it does cause a slight darkening of wetted flour as measured by the Agtron colormeter, a test that is used as an index of milling refinement and not visual acceptability.

3. SEPARATION

Reduced iron is of such a fine particle size that while it has a much higher density than wheat flour, its surface area: weight ratio is similar or even higher. If reduced iron is added to granular materials, such as farina or semolina, it can be removed along with the dust when run through a purifier or any similar device that uses air separation. This problem can be prevented by addition of reduced iron at the end of the milling process.

Reduced iron is magnetic and may be picked up on magnetic separators used to remove tramp iron (Anonymous, 1972). The flow of the product past the magnet will continually dislodge iron. Since it may fall off in concentrated amounts, some form of reblending is necessary after the magnet to ensure that no segregation of the iron remains. This may be something as simple as a screw conveyer. In one test at a bakery using magnetic separators on incoming flour, no significant difference in the iron content of the bread produced was found (Fortmann *et al.*, 1974). The mixing of large quantities of dough probably nullified any iron segregation that may have occurred.

There has been concern that the high density of reduced iron may cause segregation problems. Such problems have not been observed with flour. The iron, apparently, is unable to migrate through flour or any other cereal product, like cornmeal, containing a significant proportion of "fines" (Ponte, 1979). There may be a problem with granular products containing little or no fines, like semolina and farina.

Certain baking processes may lend themselves to iron separation problems. One study (Jaska and Redfern, 1975) showed a lowered iron content in breads made with a continuous-mix process using 20% or less flour in the brew sponge when the flour was fortified with reduced iron but not when fortified with ferric orthophosphate. This was explained by the greater specific gravity of reduced iron compared to that of ferric orthophosphate, which prevented the former from being properly suspended in the less viscous flour brew containing 0 to 20% flour. Brew flour levels of at least 25% were needed to prevent the iron from settling to the bottom of the tank.

4. COMMERCIAL AVAILABILITY

There would be little point in basing an iron fortification program on a particular iron source that could not be obtained in the necessary quantities or could not be produced to the desired specification. For example, there have been reports of experimental samples of ferric orthophosphate and sodium iron pyrophosphate showing acceptable bioavailability. Such samples are generally of an extremely fine particle size and do not represent commercially available products. Neither does hydrogen-reduced iron with a particle size less than 10 μm, whose bioavailability has been shown to approach that of ferrous sulfate (Shah and Delonje, 1973). While it may be possible to make such products, no one is currently doing so.

Whatever iron source is used, it should be a food grade material, meeting those specifications given in the *Food Chemicals Codex* or similar publications, to ensure its purity and lack of excessive contamination with heavy metals.

5. COST

After bioavailability and functionality problems are recognized, if not necessarily solved, the final arbiter determining the source of iron to use may come down to cost. Cost comparisons of different iron compounds, as given in Table IV, must take into account the percentage iron contained in each source. These figures represent 1980 prices in U.S. dollars for large quantities of undelivered product. Actual costs would be somewhat higher.

In cost per quantity of iron the least expensive source is reduced iron. This is followed by electrolytically reduced iron and ferrous sulfate, which are generally comparable. The 1978 prices for these two had ferrous sulfate slightly cheaper, but here it is reversed. It would be nice to be able to compare these sources in cost per quantity of "available" iron if one knew what relative biological values (RBV) to assign to each. The RBV of hydrogen-reduced iron would need to be

Table IV
RELATIVE COSTS OF COMMONLY USED IRON SOURCES

Iron source	Cost[a] (U.S. $/kg)	Contained iron (%)	Cost of iron (U.S. $/kg of Fe)
Ferrous sulfate, dried	1.43	32.1	4.45
Hydrogen-reduced	1.06	97.0	1.09
Electrolytically reduced	3.20	98.0	3.26
Carbonyl-reduced	6.37	98.0	6.50
Ferric orthophosphate	1.70	28.6	5.94
Sodium iron pyrophosphate	1.94	15.7	12.36
Ferrous fumarate	3.28	33.0	9.95

[a] 1980, Large volume undelivered.

less than 24% in order for this cost index to be higher than ferrous sulfate (whose RBV is 100 by definition). Since the RBV of hydrogen-reduced iron is likely higher than that, it probably still has the least cost per quantity of available iron of any iron source.

In terms of meeting dietary needs, the cost of the iron source itself is not the problem. One cent worth of ferrous sulfate will supply a person's entire dietary requirement for iron (18 mg/day) for a year. The real cost comes in distributing the iron to those people who need it.

B. Iron Levels

Iron enrichment standards are normally set by the government. These standards may be based on one of two types of rationales: restoring to flour the iron removed by milling or using flour as a vehicle for providing additional iron to the population. In either case there should be evidence of deficiency, or potential risk of deficiency, within the population. The amount of iron added should be enough to have a beneficial effect on the health of the population. But it should not be so high as to cause problems in product quality.

Most countries have standards based on restoration levels, that is, in the 29- to 36-ppm range. Great Britian has a lower standard (16.5 ppm). It is based on restoring the iron content back to that contained in an 80% extraction flour rather than to whole-wheat levels. The United States increased its iron enrichment standard for flour in 1982 from 29–36 ppm to a minimum of 44 ppm (Federal Register, 1981). This was done to make it easier to use enriched flour to meet the iron standard for enriched bread (set at 27.5 ppm), assuming that bread contains 63% flour.

These iron enrichment standards are the levels of iron that need be present in flour. The amount of iron actually added is somewhat less, since it need only make up for the difference between the enrichment standard and the natural iron content of the flour while providing a safety overage to ensure that the minimum requirement will be achieved (usually 10%). The amount of iron added to achieve a 29- to 36-ppm standard is generally 25 ppm. Under the new minimum standard of 44 ppm in the United States, the amount of iron that needs to be added is 37 ppm.

A true iron fortification program for wheat flour would be one that provides a higher amount of iron in fortified flour than that normally present in whole wheat. Additional iron would be added to cope more effectively with a severe iron deficiency problem than what might be accomplished through a standard enrichment program alone. This was once proposed for the United States (Federal Register, 1970). The standard was to be 88 ppm. This was never used for a variety of reasons (Federal Register, 1977), but it was demonstrated to be technically feasible.

The level of iron to be added is a function of the extent to which the target population is deficient in their iron intake and the flour consumption of that population. As an example, suppose it has been determined that a population with an average flour consumption of 50 kg/person/year needs, on the average, an additional 5 mg of iron per person per day. The amount of iron that needs to be added to flour to accomplish that is 36.5 ppm. This amount of iron would be added regardless of the natural iron content of the flour. A 75% extraction flour having 11 ppm iron would then contain 48.5 ppm. Higher extraction flours would contain from 50 to 70 ppm depending on their indigenous iron content.

While such true iron-fortification programs have been proposed, none have actually gone into effect. One reason for this is the difficulty in accessing the distribution of iron intakes within a given population and relating that to flour consumption patterns. In most cases the data are simply not available. It is far easier, as most countries have done, to base the fortification program on restoration. This is unfortunate, since the cost of adding additional iron over that needed for restoration would be small compared to the potential benefits.

It might be advisable to increase the level of iron added to account for differences in iron bioavailability. For example, an iron enrichment standard of 30 ppm based on ferrous sulfate might be set. This could be achieved by adding that much ferrous sulfate or twice as much reduced iron, assuming the reduced iron has a relative biological value of 50%. Such a program could be as effective yet cheaper than a program requiring one of the more expensive sources of iron.

C. Vitamin–Mineral Premixes

Since the fortification of wheat flour usually involves more than one nutrient, it is common practice to add all of the nutrients in the form of a single "premix." A typical formulation for a fortification premix includes, along with the nutrient sources, a filler (which may be cornstarch, wheat starch, or calcium sulfate) and a small amount of a free-flowing agent such as tricalcium phosphate. The premix is made by blending all the ingredients together in any suitable mixer such as a ribbon blender. The premix is then analyzed for the nutrient components to ensure their presence in the proper concentrations.

Commercially prepared premixes of certified nutrient content are available from a number of different manufacturers. Premixes can also be prepared on site. The simplest method of mixing is to use a 55-gallon drum. The drum is filled with the ingredients, to no more than two-thirds full, and mixing is accomplished by rolling the drum back and forth. This type of mixing is not very satisfactory for those nutrients, like niacin, that tend to clump.

Most nutrients, including all of the iron sources, are very stable in such a premix. The premix can be kept for a year or more before being used as long as it is kept sealed and not exposed to high temperatures or humidity.

A good fortification premix should have a bulk density close to the product being fortified. It should be relatively free-flowing but not overly so, and should not cake on storage. It should not have an excessive moisture content.

The cost of any premix, in addition to the direct cost of the nutrient sources, includes the cost of manufacturing, quality control, sales, and distribution. Import duties, taxes, clearing, and forwarding charges may also be involved. The final cost of addition at the mill will be two to three times the cost of the iron source with these expenses included.

D. Methods of Addition

1. MILL ADDITION

It is a rather simple process to add an iron source or fortification premix in a batch process. All that is needed is a small scale to ensure it is dispensed with sufficient accuracy. Wheat flour, however, is normally produced by a continuous process requiring that the nutrients be constantly metered into the flow of flour. This is accomplished with a volumetric or screw-type feeder, as shown in Figs. 1 and 2. There are several brands of these types of feeders available. Years of experience have proved them to be highly dependable and accurate enough (within 4%) for use in flour fortification.

Sufficient premix can be placed in the hopper to allow the feeder to run unattended for a number of hours. The feed rate can be manually adjusted to match the rate of flour flow. There are feeders available that automatically make this adjustment.

The fortification premix can be either fed directly into the flour by gravity or air-conveyed. In a gravity system the feeder is placed over the input end of a flour conveyor, preferably a long "cut flight" or mixing-type screw conveyor to

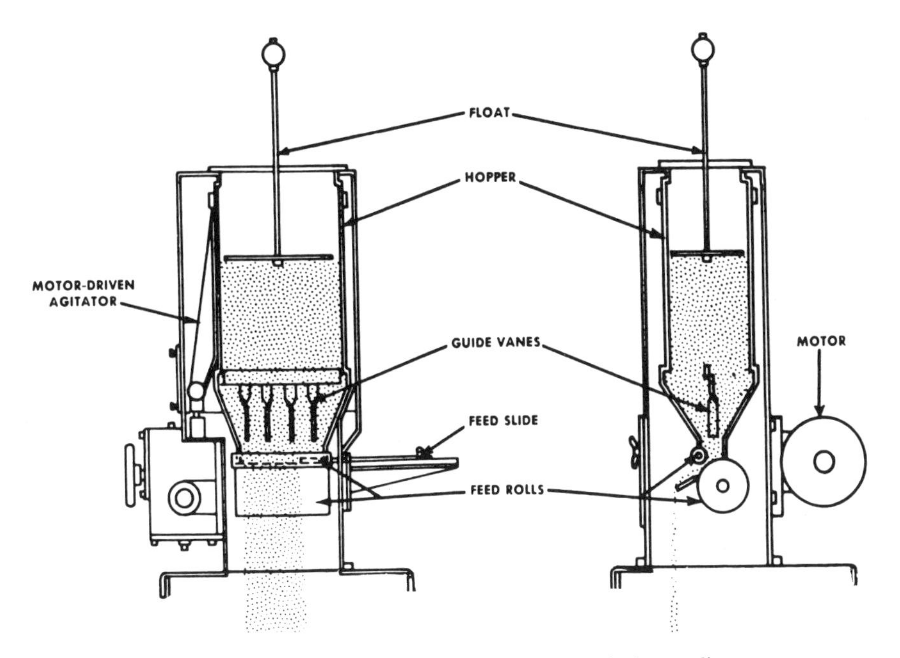

Figure 1. Roll-type volumetric feeder with slide bar feed-rate adjustment.

ensure adequate blending. A pneumatic system allows the feeders to be placed in a remote, centralized control location. This system requires an air blower, an "ejector" unit, and pneumatic lines to carry the nutrients to the point of application, which can be a screw conveyor, agitator, or entolator. It is very important that some degree of mixing be provided after the point of application, since a small amount of fortification premix is being added to a large amount of flour; typical application rates for premixes run from 0.1 to 0.3 g/kg of flour.

The 1980 cost of this equipment (in U.S. dollars) is around $2000 for a feeder, $1000 for a blower, and $500 for associated equipment and installation fees, for a total cost of $3500.

Correct adjustment, frequent monitoring, and preventive maintenance are required for proper functioning of a flour fortification system. The feeder is adjusted on the basis of the stated application rate of the premix and the throughput of the flour at the point of application. As an example, if the premix is to be fed at a rate of 0.25 ounce per hundred weight (oz/cwt) and the mill is producing 100 cwt/hour, the feeder should be adjusted to add 11.8 g/min:

$$(0.25 \text{ oz/cwt})\ (28.35 \text{ g/oz})\ (100 \text{ cwt/hour})\ (60 \text{ min/hour}) = 11.8 \text{ g/min}$$

One of the biggest problems in adjusting a feeder is knowing the correct flour throughput, which may not only change after the feeder is set but also vary

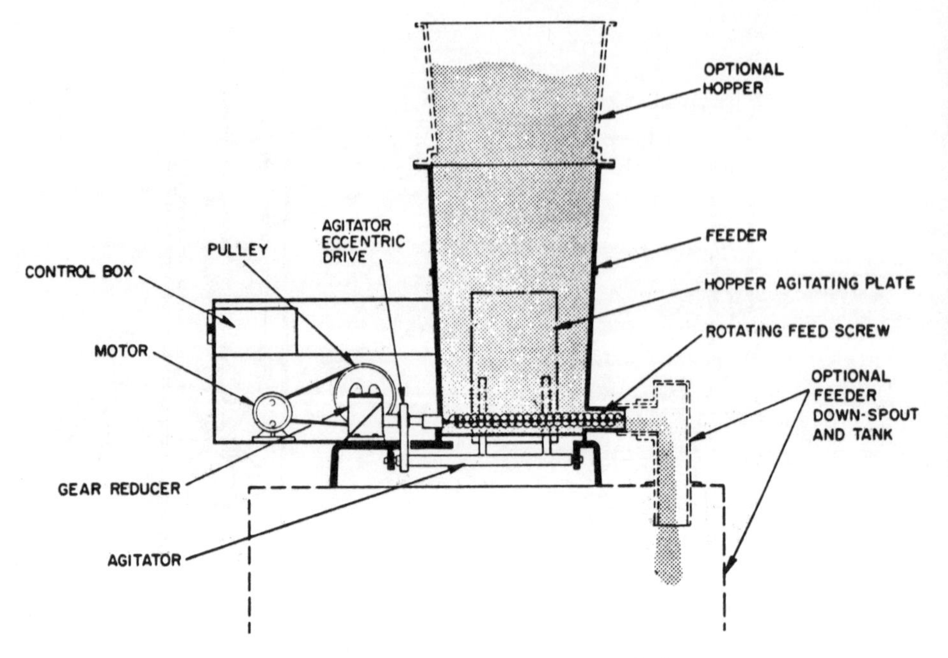

Figure 2. Variable-speed-drive screw-type feeder.

throughout the day. A feeder should be given a couple of 1-min check weights during each day's operation and, when necessary, readjusted in line with any change in the rate of flour production. The use of automatically adjusting feeders ensures that the addition rate is correct despite changes in flour production. A system of record keeping that shows the disappearance of fortification premix in relationship to the total flour production is highly desirable. It is possible to place the feeder on a scale or some other weight-recording device allowing for a continuous record of disappearance. Such disappearance records can be used by government regulatory agencies as one way to ensure the flour is being fortified.

While most feeders can run for months on end without requiring much maintenance outside of greasing and oiling, they should be checked. When they eventually wear out they need to be replaced or rebuilt. Pneumatic lines and ejector throats can also be clogged with a buildup of the premix. They then need to be cleaned (by soaking in water) or replaced.

Iron fortification of flour should be monitored at the mill by running qualitative spot tests for iron (see Section VIII, E) or, in the case of very large mills, quantitative tests on the final product on a regular basis.

In summary, a successful system for fortifying wheat flour with iron requires the following:

- A well-made, quality fortification premix
- A dependable, accurate feeder
- A proven point of application
- A preventive maintenance program for the feeder
- A system of records reflecting premix disappearance
- A system of frequent qualitative monitoring
- A means of regular quantitative monitoring

Where wheat flour is producted by a few large, centrally located mills, it can be efficiently and economically fortified. But where it is produced by a large number of small village stone mills, fortification becomes much more difficult. One suggestion for such mills is to use a block of compressed nutrients pressed up against the stone so that they are continually discharged by abrasion. This idea has not been tested, so its practicality has yet to be determined.

2. BAKERY ADDITION

Since baking is a batch process, rather than a continuous one like milling, the enrichment methods are somewhat different. It is possible to add the same powdered premix, or iron salt, used in milling. This is the cheapest product to use, but it requires the most labor and is prone to error because a separate weighing is needed for each batch of dough. Prescaled products, formulated for specific sizes of doughs, are more often used. Compressed tablets and wafers are normally designed to be added one for every 100 lb of flour used. They are scored in quarters so they can be easily broken for addition to smaller doughs using 50 or 25 lb of flour. Another prescaling method is packaging the premix or iron salt in a capsule or pack that dissolves when water is added.

Sometimes the baker forgets to add the tablet or adds it twice. The occasional over- or underfortification that results is undesirable for economic, nutritional, and consumer acceptance reasons. One way of preventing this is to combine the nutrients with some other bakery ingredient or additive, without which a poor-quality product would result. The most obvious ingredient to do this with is the flour. Other ingredients that can be fortified include yeast and salt, which work quite well, and sugar, which has not proved very successful. The trouble with this method is that the baker is locked into using a set level of the fortified ingredient. The yeast and salt levels stay fairly constant, but there are still times when the baker would like to change them, whereas the sugar level can show wide variation. The iron can also be combined with additives like potassium bromate, enzymes, or dough conditioners. Since these are used in small amounts, they can be combined in one tablet or soluble package.

E. Quality Control

Any iron fortification program requires periodic analyses to ensure the iron is in the food in the desired amounts. The best devised regulations can be ineffective if they are not accompanied by proper means of control and some degree of enforcement. Facilities and procedures for assaying the iron content of foods along with a properly trained technical staff are needed for this purpose.

1. QUALITATIVE SPOT TEST

A simple spot test, American Association of Cereal Chemists (AACC) (1983) method 40-40, can be used to determine if iron has been added to wheat flour or other ground cereal products. A 10% KSCN solution is mixed with an equal volume of 2 *N* HCl just before use. A few drops of this reagent, sufficient to wet an area about 1 cm in diameter, are applied to a flour slick, followed by a few drops of 3% hydrogen peroxide. After about 10 min, small red spots will appear if reduced iron is present. If ferrous sulfate is present, large red spots rapidly appear. The density of the spots affords some estimate on the level of iron fortification. It is best to use this method in comparison with samples of known iron content.

2. QUANTITATIVE METHODS

All analytical procedures for determining iron in food products consist of two separate steps: extraction and detection.

a. Extraction. No procedure will be of value if the iron cannot first be extracted from the test samples. There are two procedures in current use: dry- and wet-ashing. Dry-ashing methods, as given by the Association of Official Analytical Chemists and the American Association of Cereal Chemists (AACC), basically call for ashing in a muffle furnace at 500–600°C overnight. This is followed by addition of concentrated HCl, evaporation to dryness, and solution in dilute HCl. While dry-ashing is the most simple to do, there are problems with sample loss, volatilization, and iron contamination.

In wet-ashing the sample is hydrolyzed with concentrated acids at high temperature and/or pressure. While this method provides a complete extraction, there is a danger in handling the hot, concentrated, oxidizing acids required.

b. Detection. Once the iron has been extracted from the sample, there are a number of detection methods available. The most commonly used are those based on colorimetric and atomic absorption principles.

Colorimetric procedures, as in AACC (1983) procedure 40-41A, involve reducing the iron to the ferrous form with a suitable reducing agent (such as hydroxylamine hydrochloride or ascorbic acid), adjusting the pH, and reacting with an appropriate color agent (such as α,α-dipyridyl or orthophenanthroline). The test can be run manually or by automated chemistry methods.

Atomic absorption, as in AACC (1983) method 40-70, is more costly to use but allows the detection of a number of different minerals from a single extract. A number of instruments are available that are suitable for this purpose.

While many countries regulate the amount of iron in flour and bread, there is very little regulation of the type of iron. The United States requires only that the form of iron used be "harmless and assimilable," but it does not define what either of these mean. Canada has a new regulation (Canada Gazette, 1983) that reduced iron used in enrichment have a particle size so that 100% will pass through a 100-mesh sieve and at least 95% will pass through 325-mesh (44 μm). If sodium iron pyrophosphate is used, it must have a bioavailability not less than 50% that of ferrous sulfate.

IX. PASTA PRODUCTS

Pasta products (macaroni, spaghetti, noodles, etc.) are popular foods in many countries and represent an important part of the diet. They are especially important in regard to iron nutrition when used as a meat substitute, as they often are when meat gets very expensive.

Pasta products are made from milled, refined wheat (usually durum wheat). Unlike flour and bread, whole-wheat or high-extraction pasta products were never in common use, nor are they much in use today. For this reason, pasta has never been a good source of iron and only becomes one if fortified.

In those countries that fortify pasta products with iron, such as the United States, it is usually supplemented up to the same iron level as that used for flour (29–36 ppm). In Canada the level of iron fortification is not regulated and varies from company to company.

The source of iron used has traditionally been the poorly available ferric orthophosphate or sodium iron pyrophosphate. Reduced iron causes undesirable spots in pasta and cannot be used. It was long thought that ferrous sulfate would cause the same problems in pasta as it does in wheat flour, especially since pasta products are required to have a long shelf life. It has since been discovered that ferrous sulfate can be used in pasta without adversely affecting its color, shelf life, or flavor. Nearly all of the pasta produced in the United States is now being fortified with ferrous sulfate (Ranum and Loewe, 1978). The reason that ferrous sulfate can be successfully used in pasta fortification has not been well documented. Because of its high bioavailability, it would be the preferable iron source to use but may need to be tested for use in unusual types of pasta products to ensure that there are no functional problems.

Pasta fortification can be performed at the pasta plant or by adding the nutrients at the mill to the semolina, durum flour, or wheat flour from which the pasta is made. It is often more convenient to do the latter, since there are many more pasta plants than there are mills.

X. CEREAL IRON ENRICHMENT PROGRAMS

A. Rationale

The general term *fortification* applies to any public health measure aimed at improving and maintaining the health of individuals in a population through the addition of adequate levels of nutrients, to foods. *Enrichment* describes one type of fortification in which nutrients are added back to a food to those levels it previously contained before they were removed by processing. Enrichment is often thought of as a preventive program aimed more at maintaining adequate nutrient intakes rather than improving them. The levels of nutrients provided by an enrichment program may be insufficient when serious nutritional deficiencies exist. In that case, it may be desirable to go to a true fortification program in which higher than normal levels of the nutrient are added or nutrients are added to foods that normally do not contain them, as when iron has been added experimentally to salt and sugar.

Effective fortification programs can only be achieved through government action. The individuals that make up a country's population cannot be expected to pay attention to a condition like iron deficiency anemia until it develops into a disease state. At that point they may seek medical help and be clinically treated, but if enough individuals in the country are in this condition, the country as a whole suffers. In that case, the government has a clear responsibility to provide suitable prevention programs.

Two reasons are traditionally given for fortifying milled cereals (wheat, corn, rice, sorghum) with iron or any other nutrients. The first is to restore those nutrients removed by milling. The second is to use these important food staples as vehicles for getting the nutrient to population groups who are known to be deficient or who have a potential risk of nutrient deficiency. Furthermore, cereals are usually inexpensive basic food staples consumed by many people and therefore offer an economically feasible means of nutrient delivery. There is a clear need to increase the dietary iron intake of many, if not most populations, and cereal fortification is a potentially feasible route.

A basic premise behind the replacement of nutrients removed by processing is that the food should be made to provide the same level of nutrients to the diet as it formally did before the processing became commonplace. The many small, coummunity-located stone mills producing a high-extraction flour have been replaced by a few high-volume, highly efficient, centrally located roller mills producing a lower extraction flour with reduced iron content. A similar trend has also occurred with baking. Bread and other breadstuffs are manufactured products, more of which are bought across the counter than are made at home. The establishment of large commercial bakeries has proceeded side by side with that of modern wheat mills.

An important trend in wheat consumption is the growing demand for breadstuffs other than bread. Cakes, pastries, breakfast cereals, biscuits, cookies, crackers, bulgur, and a host of other products, all based on wheat or flour, are becoming increasingly popular throughout the world.

These trends, which started in the industrialized countries a century ago, are now affecting diets in developing countries. In developing wheat-growing countries the share of the total flour production contributed by village stone mills is continually lessened with each modern milling facility constructed. In the traditional rice-eating countries of South and East Asia and corn-eating countries of Latin America, whatever flour was needed was imported as flour. As the demand for wheat products expanded, large flour mills were established at seaports to process imported wheat.

One result of this centralized industrialization of milling is a slow decrease in the iron content of the cereal product and potentially the overall diet. One estimate on the extent of this decrease in Norway (Hallberg *et al.*, 1979) showed a daily iron intake of 7.2 mg/1000 kcal in the thirteenth century, 80% of which was derived from barley. In the nineteenth century, this dropped to 6.1 mg/1000 kcal, of which 60% came from wheat. The daily intake would now lie between 4 and 5 mg/1000 kcal were there no iron fortification of flour and bread. The reasons given for this decline were (1) lowered intakes of foods rich in iron and high intakes of foods low in iron, (2) less iron contamination of foodstuffs picked up from cooking pots, utensils, and machines than in the past as iron has been replaced by aluminum, stainless steel, and plastics, and (3) foods that could deliver the most iron, namely the cereal grains, having lost some iron through the milling process.

It has been argued by many that we should go back to the high extraction rates of yesteryear so as to retain as much of the indigenous nutrients in cereals as possible. This is a highly controversial subject about which much has been written (Dunlap, 1945); it cannot be adequately dealt with here. Suffice it to say that it is not likely to happen on a worldwide basis. A high proportion of the world's flour will continue to be produced by extraction rates in the 70–80% range. When such flour becomes an important part of a national diet and when there is a demonstrated need for more iron in the diet, the addition of iron to the flour is a more practical and cost-effective measure than trying to force a fundamental change in a country's milling and baking technology.

Countries that require that iron be added to wheat flour include Canada (29–43 ppm), Chile (30 ppm added), Denmark (30 ppm), Guyana, Kenya, Zambia (all 29–36 ppm), Great Britain (16.5 ppm), Nigeria (35 ppm), and the United States (44 ppm).

B. Effectiveness

A limited number of studies have been done to determine the effectiveness of iron fortification toward improving iron intakes. Three studies (Hallberg *et al.*, 1979; Cook *et al.*, 1973; Bazzano and Carter, 1978) involved the addition of iron sources to flour and bread. The most recent and complete study (Hallberg *et al.*, 1979) was done in Sweden. It showed that an increase in the iron fortification of wheat flour accounted for about 40% of the improvement in the iron status of Swedish women. The iron deficiency rate in Sweden has been reduced from about 25–30% that was present when iron fortification was first started, to the present level of about 7%. The daily amount of added iron is 8 mg. This is coupled with the native iron content of the diet of 11 mg to make a daily intake of 19 mg of iron. Reduced carbonyl iron is the iron source. Other dietary factors such as the use of iron and ascorbic acid supplements also have contributed to the improvement.

C. Consideration of a Possible Program for the Foritification of Wheat Flour with Iron in Egypt

Egypt is a country with a serious iron deficiency anemia problem. It does not currently have an iron fortification program, but it is considering establishing one. Following is a review of the feasibility of fortifying wheat flour with iron in Egypt.

Anemia has been demonstrated to be a major public health problem in Egypt. In 1974, a report in the *WHO Chronicle* (Rao, 1974) that included the status of malnutrition in Egypt showed iron deficiency anemia as being widespread and a serious problem. It estimated that anemia affected 20–25% of the children, 20–40% of adult women, and 10% of the adult men.

The National Nutrition Survey of 1978 (Arab Republic of Egypt, 1978) and the Nutrition Status Survey II carried out in 1980 (Arab Republic of Egypt, 1980), both conducted by the National Institute of Nutrition, Ministry of Health, Arab Republic of Egypt, in cooperation with AID and CDC (Center for Disease Control, USDHHS), confirmed that high prevalence of anemia. The data from these studies estimated that 1.4 million children are anemic. The studies suggested that it is primarily a result of insufficient bioavailable iron in the diet. Mothers of the children surveyed also were studied, and the results indicated that anemia is widely prevalent in women of childbearing age, especially those pregnant and lactating.

Although the general nutrition of the population has improved, there appears to have been no improvement in the iron deficiency anemia status of the people since that reported in 1974. The government of Egypt recognizes the size and importance of the iron deficiency anemia problem. The result of their own studies coupled with the reconfirming results from other studies encourage the government of Egypt to carry out programs for combatting anemia. Various alternatives or types of interventions are available that can form part of a strategy for combatting iron deficiency anemia. Those alternatives include programs of supplementation, food fortification, and weaning food development along with agricultural production and nutrition education.

1. VEHICLE

The most important staple food in the diet of the Egyptian people is wheat. Wheat in the form of flour or the products made from it such as Balady and Shamy breads and pasta comprises nearly 50% of the daily intake of food and represents approximately 30–40% of the total calories and protein consumed in the Egyptian diet.

Wheat flour and bread are the principal subsidized foods. The demand for wheat has grown nearly 75% in the last decade. Imports have doubled to keep up with demand. Domestic wheat production has decreased or at least not kept pace with demand. More than 70% of the wheat consumed is imported and milled into flour in Egypt.

There are 150 government mills, of which 103 are stone mills and 47 roller mills. These mills process about 5.5 million metric tons of wheat per year into flour, which is approximately 90–95% of the flour produced in the country. The stone mills produce about 60% and the roller mills about 40% of the total flour produced. Several 500-ton roller mills are being built or are planned for construction before the year 2000. As roller mills are built, the stone mills are closed.

The flour produced in Egypt is 82% extraction, that is, it contains 82% of the whole-wheat kernel. This type of flour is used to make traditional Balady bread. The uniformity and quality of the flour and the control of its production is much better in the roller mills. The Ministry of Supply controls the production and use of the flour.

The private-sector mills are reported to process 500,000 metric tons of wheat per year. This represents about 7% of the flour consumed in the country. There are about 33 nongovernmental mills.

In addition to the flour produced in the country, 1,320,000 tons of flour are imported each year. This flour is of 72% extraction. About 65% of this flour is used in commercial bakeries to make Shamy (a form of Balady bread), and French bread, and for the production of pasta. The remainder of this flour is used in the home, military, and other institutions primarily for bread production. The use of imported flour also is controlled by the Ministry of Supply.

More than 90% of the wheat flour, whether produced in-country or imported, is used to produce bread of one type or another. There are innumerable bakeries. About 15% of the bread produced is from government bakeries operated by the Ministry of Supply. These bakeries produce Shamy bread and/or French bread using semiautomatic and automatic equipment. Some of these bakeries also produce Balady bread by the usual hand methods. The remaining 85% of the bread is produced in small bakeries that use hand methods. This bread is all Balady bread. The subsidy on bread has discouraged home baking, so that essentially all bread consumed is made from flour produced in commercial mills.

The Ministry of Supply reports the average yearly consumption of wheat as 175 kg per person. It is described as being consumed as an average daily consumption of three loaves of Balady bread or four loaves of Shamy bread or French bread. The official weights per loaf of these breads are 169 g for Balady and 125 g for Shamy and French.

Unpublished data from an AID-supported project in Egypt on "The Functional Significance of Marginal Nutrient Deficiencies" shows the actual consumption of bread per day by members of the families in their survey as 565 g for adult male, 480 g for adult females, 318 g for school-age children, and only 41 g for preschoolers. The data in the first three categories bear out the reported "average" daily consumption of bread, as it is within the range of two to four loaves of Balady bread per day at the official weight. It was reported that the loaves of Balady in the village weighed about 230–250 g, as they were mostly prepared in the home. However, it is the agreement in total consumption (weight of bread or flour consumed) that is of most concern and that would be used for later calculations regarding iron fortification. The total consumption of bread can be equated to a given quantity of flour. It is this figure that would be the basis for the addition of iron to flour in a food fortification program.

The reduced consumption of bread by preschoolers indicates that while everyone over the age of 2 years eats bread, many preschoolers would not get significant portions of their daily iron requirement from eating bread made from flour fortified with iron at levels that would be significant for other members of the society.

Wheat flour would be a good vehicle for carrying deficit iron to the Egyptian people since it is consumed in large quantities by all but the youngest children. Wheat flour is centrally milled. This offers a good point of intervention for adding iron to it while it is being produced. Adding the iron at this point using the technology developed for the iron fortification of wheat flour would ensure the presence of added iron in all flour, thus all bread and other foods containing wheat flour. Also, it would assure a high degree of compliance in the regularity and accuracy of the addition of the iron.

Concurrent with such an intervention would need to be a requirement that all imported wheat flour contain added iron at a specified level.

The technological method for fortifying wheat flour with iron is well developed and has been used for many years in many countries. Presently it is being used in the production of fortified flour in more than 20 countries, both developed and developing. At least a dozen countries that do not produce their own wheat flour require that the imported flour be fortified with iron as well as certain vitamins in many cases. Fortification of wheat flour with iron is a simple, effective, inexpensive way of introducing bioavailable iron into the daily diet.

2. GENERAL TECHNOLOGY FOR IRON FORTIFICATION OF WHEAT FLOUR

Flour fortification technology involves three major components: (1) the feeder, (2) the composition of the iron source of iron-containing material to be added (premix), and (3) the control of the rate of addition of the premix. The feeder normally used is a dry-material feeder capable of accurately adding small quantities of material such as a premix containing iron. The quantity of addition generally is from 15 to 31 g premix per 100 kg flour.

One feeder per flour mill is all that is required. It is installed in the mill at the proper place so as to add the premix to the flour in the last stage of processing just before packaging. Installation and operation of the feeder is simple and can be done by any general flour mill personnel. Manufacturers and sellers of the feeders often will provide guidance and assistance in the installation and initial use of the feeders.

The iron source to be added to the flour is mixed with a small amount of carrier (usually wheat flour or starch). This blend is called a premix. It is used to facilitate the ease and accuracy of the addition of the iron source through the feeder. The composition of the premix is based on the form of iron to be used and the amount of iron desired to be added per kilogram of flour. Recent data for the United States show that in the iron fortification of cereal foods, 45–50% is done with reduced iron, 40–45% with ferrous sulfate, and less than 10% with other forms of iron. Ferrous sulfate and reduced iron are used because of their favorable properties of bioavailability of the iron, relative stability, and low cost.

The amount of iron source to be contained in the premix is determined by the desired iron level in the flour and the proposed application rate. In the United States, nutrient premixes are formulated so that an application rate of 15–31 g/100 kg will add 38 mg iron/kg flour. This level of addition meets the standards for the iron fortification of wheat flour in the United States. Premixes are formulated so that the application rate can be held as low as possible to minimize the amount of premix introduced into the food being fortified.

The cost of initiating and sustaining a program for the iron fortification of wheat flour is composed of (1) capital costs associated with the equipment and activities necessary to start the program and (2) the operating costs for nutrients, depreciation, and similar expenses to maintain the operation of the program.

The capital costs are one-time costs and are primarily the cost of purchasing the feeders for adding the iron premix plus any expense of installation of the feeders in the flour mills. The current cost of an appropriate feeder is about U.S. $3500 per unit.

The operating costs are those for purchasing the iron source, preparing the premix, and distributing the premix to the mills. The present cost of ferrous sulfate sufficient to add 38 mg iron/kg, flour is U.S. $0.00015, or for 100 kg flour, U.S. $0.015.

Individual private-sector companies manufacture nutrient premixes and make additional charges for blending and delivery in addition to the cost of the iron source. A typical cost for treating 100 kg of flour with the level of iron stated before using a manufactured premix delivered to the flour mill is U.S. $0.044.

Each mill could have a feeder installed above a conveyor that transports the flour from the last steps of sifting to the packing area where it is put into bags. This is a common practice in roller mills. Feeders can be adequately installed in the stone mills, also. This will assure the proper addition and adequate blending of the premix with the flour before it is packed.

There are approximately 4900 bakeries of record under the ownership or supervision of the Ministry of Supply. It would present a formidable, almost insurmountable task to attempt to distribute an iron-containing premix to each bakery and expect it to be regularly and accurately used. Only about 30% of these bakeries are equipped with any degree of automatic equipment that lends itself to the control and accuracy needed for the proper addition of nutrient premixes.

The other bakeries are ‘‘hand’’ bakeries in the full sense. Their inherent method of operations renders it infeasible for these bakeries to be used as the intervention point for adding iron into the daily diet of the Egyptian people.

3. ECONOMICS

As mentioned earlier, there are capital costs and operating costs associated with the initiation and continuation of a fortification program. The one-time capital cost to install a feeder in each of the 150 governmental mills would be approximately U.S. $525,000. When prorated across the 42 million people who would be the beneficiaries of the iron fortification of wheat flour, this represents an initial expenditure of U.S. $0.013 or a little more than 1 cent per person. The operational life of a typical feeder can be expected to exceed 30 years.

The operating or recurring costs are almost all in the cost of the iron source to be added to the wheat flour. The following cost analysis for the addition of iron as ferrous sulfate is based on providing a level of added iron of 9 mg/person/day for a year to each of the 42 million people of Egypt.

Ferrous sulfate is the iron source of preference when the properties of the vehicle (food) and the conditions it encounters in use allow it to be used without difficulty.

Ferrous sulfate is not recommended for use in flour that has a lengthy storage or transportation period between the time it is milled and when it is used. Also, it generally is not used in flour that is stored or used under hot and humid conditions. The conditions existing in Egypt at the present time result in the flour being used within a period of weeks after its production, with essentially no storage and with short lines of distribution. Although high temperatures are encountered, it is reported that humidity is not of concern. Thus, the use of ferrous sulfate appears appropriate for products with short-term storage.

The level of 9 mg was arbitrarily selected for this analysis because of its relation to the various values in the table of Recommended Daily Allowances (USRDA). The 9-mg level equals 50% of the USRDA of 18 mg for females (not pregnant or lactating) and for males aged 11–18. Also, this level essentially equals the USRDA of 10 mg for adult males and for children ages 3–10.

The current native iron content of the diet of the Egyptian people is placed at 9 mg/day. The proposed 9-mg addition is a 100% increase in dietary iron and would reach the USRDA for essentially all but pregnant and lactating women.

Based on the per capita consumption of flour, which is about 430 g/day, the iron could be added at a rate of 20 ppm, which would provide the desired additional daily intake of 9 mg of iron. This level of addition of iron is appreciably less than that presently used in the United States and most other countries in which iron fortification programs are under way.

The cost of making available to the flour mills sufficient ferrous sulfate in a premix to provide the aforementioned increase in daily intake of iron would be approximately U.S. $672,000 per year or U.S. $0.015 per person per year. This cost per year per person is about equivalent to the subsidy associated with the price of one loaf of Balady bread. When compared in this manner, a program for the iron fortification of wheat flour could be deemed to be economically feasible.

To put these costs in another perspective, compare them to the cost of providing iron via iron supplementation tablets as is currently done in Egypt. An iron tablet contains 200 mg ferrous sulfate, which contains 60 mg iron. They cost £1.90* or essentially U.S. $2.00 per 1000 tablets. It would require consuming 55 tablets to get the same quantity of iron per year as would be provided in a program for the iron fortification of wheat flour. The 55 tablets would have to be distributed throughout the year and monitored for the compliance of taking them approximately every 6 days. The cost of the tablets alone would be U.S. $0.11 per year with no consideration for distribution or compliance costs. A yearly cost for tablets for 42 million people would be approximately $4.5 million.

*1.90 Egyptian pounds.

4. BIOLOGICAL EFFECTIVENESS

Determination of the feasibility of biological effectivnesss is a concern in a program for adding iron to the daily diet regardless of the vehicle used or the method of administration. It is important that the added iron will be biologically available to the people and will improve their iron nutritional status, in this case help reduce or eliminate iron deficiency anemia.

Upon initial examination of the flour-milling and bread-baking industries in Egypt and comparing them to those countries where iron fortification is being carried out, there seemed to be enough similarities to indicate feasibility in this area. However, two major differences became evident upon further study that warrant careful attention and that may control the determination of feasibility of biological effectiveness. The effect(s) of these differences could determine the success of a program for the iron fortification of wheat flour.

The first difference is that 82% extraction flour is produced in the country and is used in the production of the traditional Balady bread. This higher extraction rate compared to the more usual 72–75% extraction rates worldwide could have an effect on the bioavailability of the added iron.

Another factor to be considered beside the high extraction rate of the flour produced in the country is that about 70% of this flour is produced in stone mills. This unique and different method of grinding wheat in stone mills could change the characteristics and composition of the flour and have an unexpected effect on the bioavailability of added iron.

The second difference of concern is in the method of making Balady bread. It is different in that it uses a piece of starter dough from a previous batch to serve as the source of leavening. This is done in place of using bakers' yeast. The starter dough has a different flora of microorganisms, other than yeast, which are naturally present in the wheat and flour. It is possible that this type of leavening system could affect the availability of any iron that is added.

Another consideration is that additional fine bran representing about 8% of the wheat kernel becomes incorporated into the bread by virtue of its being used on the fermentation boards and in the steps of handling the dough in the preparation of Balady bread. This added bran could affect the availability of any iron present.

Since Balady bread represents over 80% of the bread consumed, any negative effect(s) that might be found on the bioavailability of iron could have nutritional implications. Thus, any effect that these differences might have should be resolved.

XI. BLENDED FOODS

The background and development of blended foods, as known today, started with the passage and amendment of the Agricultural Trade Development and

Assistance Act of 1954, referred to as Public Law 480 (PL 480, Food for Peace Program), which provides for the sale and donation of U.S. food commodities overseas. In the early days of the PL 480 Title II donation, the foods distributed through the program generally were the commodities of whole grains, ordinary grain-milling products (such as wheat flour and cornmeal), and nonfat dry milk.

These basic commodities were used individually until late 1966, when the act was amended to incorporate the use of blended foods into the program. This change was prompted by two considerations:

1. A serious shortage had developed in the domestic supply of nonfat dry milk in 1965, which emphasized the need to develop alternative foods for the PL 480 donation program.
2. Equally important, it was realized that the nonfat dry milk was the only donated food particularly suited for the diet and nutritional needs of weanling infants, preschool children, and pregnant and lactating women.

The collaborative efforts of the U.S. Department of Agriculture, the Agency for International Development, and the National Institutes of Health developed fortification guidelines for the composition of blended foods that would serve as supplementary foods for the diets of children and for the emergency feeding of adults. These foods were not designed or intended to serve as a sole source of nutrients.

In general, the nutrient levels were set so that 100 g of blended foods would supply at least one-half of the National Academy of Sciences–National Research Council Recommended Dietary Allowances for a 1 to 3-year-old child. For iron the levels set were from 5 up to 15 mg/100 g of dry blended food.

The initial blended foods developed were corn–soy–milk (CSM) and wheat–soy blend (WSB). These two foods were the mainstay of the Title II program until corn–soy blend (CSB) was introduced in 1973. The development of CSB was necessitated by a shortage of nonfat dry milk that occurred around 1971. The reformulation, however, did not alter the composition of the vitamin and mineral premixes used in the blended-food preparation. Essentially, the type and amount of iron present in the foods and its contribution to the diet remained the same.

In 1978, wheat protein concentrate blend (WPC), composed mainly of wheat protein concentrate and full-fat and defatted soy flour, was developed. It contained the same vitamins and mineral components that were used in the other blended foods.

Standard ascorbic acid (vitamin C) was used in the blended foods until 1981, when the specifications were changed to require the use of stabilized (coated) ascorbic acid in CSM, WSB, and WPC.

The spectrum of products for use in the PL 480 program was expanded between 1966 and 1970 to include not only the blended foods that were designed for use with infants, children, and adults, but also certain other nutritionally improved foods. There were the protein-fortified commodities such as soy-fortified bulgur, sorghum grits, rolled oats, and cornmeal, as well as 12% soy-fortified bread flour. However, only in the case of the soy-fortified cornmeal and the 12% soy-fortified bread flour do the final products have increased levels of iron and certain vitamins by virtue of the enrichment of the basic commodities before they are blended.

A. Types of Products

Blended foods, otherwise called formulated foods or new food mixtures, are the result of blending several component ingredients to produce a food that is nutritionally improved over any of the individual ingredients. They generally include a cereal and a high-quality protein component (soybean, nonfat dry milk, whey). Blending improves both protein quantity and quality of the product. Several dozen such foods have been developed around the world since the early 1950s. The component ingredients are mostly local, although some include imported food commodities. Those commonly used are wheat, corn, groundnut (peanut), sorghum, rice, soy, and a variety of other legumes. The foods are manufactured in central processing facilities and are distributed either through regulated feeding programs or by sale on the open market. The more well known of these products are Fortesan, Balahar, Incaparina, Faffa, Pronutro, and Thriposha. The latter four are formulated with added vitamins and minerals including iron.

The ingredients used may be milled or further processed to improve their flavor and functionality. The processing may be in the form of heating (soya, milk), rolling (oats), parboiling (bulgur,), extruding, and so on. These foods may be cooked and consumed in a manner similar to rice, or they may be made into a drink. They also may be used to extend and improve the nutritional value of other less nutritious foods, such as cassava. Blended foods often are flavored, usually with sugar, to increase their acceptability, and fortified with vitamins and minerals to improve their nutritional value.

In addition to the blended foods of the U.S. PL 480 program, many other new food mixtures have been developed around the world; such foods also are designed to alleviate malnutrition in developing countries. These foods are intended both for use in food distribution programs and for encouragement of their sale through standard retail outlets. They also are intended to become an integral part of the general diet.

In 1972 on request of the then Protein Advisory Group (PAG) of the United Nations, The Tropical Products Institute (TPI) conducted a compilation of data and review on new food blends in 36 countries so as to gain insight into their development, utilization, and benefit to recipient populations. Thirty-four distinct new food mixtures were identified, and 10 had vitamins and minerals added to them. Unfortunately, data were not available as to the quantity or type of iron added.

The Tropical Products Institute carried out a follow-up review in 1976. Their report concluded that low-income groups have limited access to new blended foods marketed through standard retail outlets. Only two blended foods, Incaparina and Pronutra, had created and sustained a notable retail trade. Both of these mixtures are fortified with vitamins and minerals including iron. Therefore, when targeting iron-fortified blended foods to the poor through retail markets, careful plans must be made to permit access of the poor to those markets.

B. Iron Fortification of Blended Foods

1. TYPES AND LEVELS OF IRON ADDED

The original guidelines and specifications for PL 480 blended foods permitted the use of iron compounds that were commercially available and acceptable as nutritional iron sources. These were studied and compared to determine which iron compound was most appropriate for use. Ferrous fumarate was selected and has since been used in the blended foods corn–soy–milk (CSM) and wheat–soy blend (WSB). Both food formulas contain 150 ppm iron.

Ferrous fumarate has a reddish brown color that often prevents its use in many basic cereal applications. However, its color is compatible with tbe yellow and brown colors characteristic of the CSM and other blended foods, and therefore is not a problem. There were no problems with solubility nor with segregation from the blended mixtures. Ferrous fumarate is chemically relatively nonreactive and therefore does not induce food deterioration as do some other more reactive forms of iron. Because of these characteristics and its reported high bioavailability, ferrous fumarate continues to be used in CSM and WSB.

Other PL 480 foods that are enriched with iron are soy-fortified wheat flour (44 ppm iron) and soy-fortified cornmeal (29–57 ppm). Reduced iron is used as the source of iron in both of these products.

2. METHODS OF ADDITION

Most blended foods are produced under batch process conditions. The dry ingredients are preblended together. The vitamin and mineral premixes are scaled separately and added individually, followed by blending of the entire batch for uniform distribution. The final step is the addition of the fat or oil component, followed by blending the batch to completion.

The protein-fortified foods that have added iron are the result of the iron, other minerals, and vitamins having been added to the cereal component during its processing. This is followed by the blending of the enriched cereal base food with the protein component.

C. Iron Contribution from Blended Foods

As indicated earlier, the products known as blended foods are designed for use in developing countries to provide, among other things, a significant source of additional iron to populations at high nutritional risk, such as pregnant and lactating women, infants, and children. The selection of ferrous fumarate was in part due to its high relative iron bioavailability. Based on this information, the amount of ferrous fumarate added to the blended foods was set at 150 ppm. This amount, coupled with the intrinsic iron present in the food, yields a total iron content of 180–200 ppm or 18 mg iron/100 g food. This is equivalent to the U.S. Recommended Daily Allowance (USRDA) for pregnant and lactating women. A 75-g serving would contain the USRDA of iron for growing children over 6 months of age. However, from studies carried out on iron bioavailability from blended foods it is estimated that infants with borderline iron deficiency would absorb between 1.1 and 2.6% of the iron contained in the blended food, or the equivalent of 0.2–0.4 mg/100 g. These data indicate that pregnant women, children, and infants are unlikely to meet their total iron needs as intended through the use of blended foods.

Many blended foods contain soy, which has been shown in some studies to inhibit iron absorption (Cook *et al.*, 1981). If the effect of soy is real, it may decrease the potential of blended foods to alleviate iron deficiency.

D. Enhancement of Iron Absorption with Ascorbic Acid (Vitamin C)

Ascorbic acid is a major enhancer of the absorption of nonheme iron. Therefore, it is a very important factor in the diets of that considerable portion of the world's population which consumes little or no meat. Other organic acids contained in some vegetables and other foodstuffs also may under certain conditions promote absorption of nonheme iron. However, the relative value of these organic acids in iron nutrition is not well defined.

The enhancing effect of ascorbic acid on the absorption of iron from maize, wheat, and soy was reported in 1973 by Sayers *et al.* This was after the formulation of the blended foods. Presumably, vitamin C initially was added to blended foods for its vitamin properties and not for its iron-enhancing effect. The blended foods had contained standard ascorbic acid that was not coated or protected, and was thus vulnerable to inactivation through oxidation, or heat or chemical reaction during many conditions of food processing and storage. However, since September 1981 (USDA Announcement CSSM-1), CSM and WSB have been formulated to contain a stabilized ethyl cellulose-coated ascorbic acid preparation. This coated ascorbic acid is believed to be much more resistant to deterioration due to moisture and temperature. Tests using the coated material under adverse conditions of temperature and moisture showed about 50% ascorbic acid activity remaining after storage for 2 months; by contrast, less than 10% ascorbic acid activity remained in uncoated test samples. The use of coated ascorbic acid should result in greater retention of ascorbic acid and greater iron absorption enhancement capability in the blended foods.

REFERENCES

American Association of Cereal Chemists (AACC). (1983). "Approved Methods." AACC, St. Paul, Minnesota.

Anonymous. (1968). "Iron in Flour," Rep. Public Health Med. Subj. No 117. H.M. Stationery Office, London.

Anonymous. (1972). AIB advises on iron enrichment bulk handling of flour. *Am. Inst. Baking, Inst. News* **29**(4), 5.

Anonymous. (1975). Problems in iron enrichment and fortification of foods. *Nutr. Rev.* **33,** 46.

Anonymous. (1983). World grain production. *Milling & Baking News* June 14, p. 63.

Arab Republic of Egypt. (1978). "National Nutrition Survey." Nutrition Institute, Ministry of Health.

Arab Republic of Egypt. (1980). "National Nutrition Survey II." Nutrition Institute, Ministry of Health.

Austin, J. (1979). "Cereal Fortification and Global Malnutrition." Oedgeschalges, Gunn & Hain, Cambridge, Massachusetts.

Aykroyd, W. R., and Doughty, J. (1970). "Wheat in Human Nutrition." Food Agric. Org., Rome.

Bassiri, A., and Nahapetian, A. (1977). Differences in concentrations and interrelationships of phytate, phosphorus, magnesium, calcium, zinc and iron in wheat varieties grown under dryland and irrigated conditions. *J. Agric. Food Chem.* **25,** 1118–1122.

Bazzano, G. S., and Carter, J. P. (1978). The Effect of Iron Fortification of Flour-based Products on Hematology in Two Study Groups," Final report to FDA, Contract No. 1 223–73–2236. Tulane Univ., New Orleans, Louisiana.

Bell, A. D. (1974). Iron enrichment of flour. U.S. Patent 3,803,292.

Bjorn-Rasmussen, E. (1974). Iron absorption from wheat bread. *Nutr. Metab.* **16,** 101.

Callender, S. T., and Warner, G. T. (1979). Iron absorption from brown bread. *Lancet* **1,** 546.

Canada Gazette. (1983). Part 1, 7703, Schedule No. 570.

Clauss, L., and Kulp, K. (1974). Abstract 76: Effect of the bromate–ferrous sulfate interaction on dough properties. *Cereal Sci. Today* **19,** 395.

Cook, J. D., Minnich, V., Moore, C. V., Rasmussen, A., Bradley, W. B., and Finch, C. A. (1973). Absorption of fortification iron in bread. *Am. J. Clin. Nutr.* **26,** 861.

Cook, J. D., Morck, T. A., and Lynch, S. R. (1981). The Inhibitory effect of soy products on non-heme iron absorption in man. *Am. J. Clin. Nutr.* **34,** 2622.

Dobbs, F. J., and Baird, I. M. (1977). Effect of wholemeal bread on iron absorption in normal people. *Br. Med. J.* **1,** 1641.

Dunlap, F. L. (1945). "White versus Brown Bread." Wallace & Tiernan Co., United States.

Edwards, C. H., Booker, L. K., Cordella, M. S., Rumph, R., and Ganapathy, A. M. (1971). Utilization of wheat by adult man. *Am. J. Clin. Nutr.* **24,** 547.

El Gindy, M. M., Lamb, C. A., and Burrell, R. C. (1957). Influence of variety, fertilizer treatment, and soil on the protein content and mineral composition of wheat, flour, and flour fractions. *Cereal Chem.* **34,** 185–195.

Elwood, P. C., Benajamin, I. T., Fry, F. A., Eakins, J. D., Brown, D. A., Dekock, P. C., and

Shah, J. U. (1979). Absorption of iron from chapatti made from wheat flour. *Am J. Clin. Nutr.* **23,** 1267.

Federal Register. (1970). Cereal flour and related products and baker products. April 1, **35,** 5412.

Federal Register. (1977). Iron fortification of flour and bread. November 18, **42,** 59513.

Federal Register. (1981). Iron fortification of flour and bread; Standards of identity. August 28, **46,** 43413.

Food and Agriculture Organization. (1970–1980). "1965 Agriculture Commodity Projections," Vol. II, p. 107. FAO, Rome.

Fortmann, K. L., Joiner, R. R., and Vidal F. D. (1974). Uniformity of enrichment in baker's flour applied at the mill. *Bakers Dig.* **48**(3), 42.

Fritz, J. C., Pla, G. W., and Rollinson, D. L. (1975). Iron for enrichment. *Baker's Dig.* **49,** 46.

Ghanbari, H. A., and Mameesh, M. S. (1971).Iron, zinc, manganese, and copper content of semidwarf wheat varieties grown under different agronomic conditions. *Cereal Chem.* **48,** 411.

Graham, R. P., Morgan, A. I., Hart, M. R., and Pence, J. W. (1968). Mechanisms of fortifying cereal grains and products. *Cereal Sci. Today* **13,** 224.

Hallberg, L. (1979). An analysis of factors leading to a reduction in iron deficiency in Swedish women. *Bull. W.H.O.* **57,** 947.

Harrison, B. N., Pla, G. W., Clark, G. A., and Fritz, J. C. (1976). Selection of iron sources for cereal enrichment. *Cereal Chem.* **53,** 78.

Jackel, S. D. (1971). Encapsulated ferrous sulfate. *Food Process.* **32**(5), 28.

Jaska, E., and Redfern, S. (1975). Interaction of ferrous sulfate with potassium bromate and iodate in brew and dough systems. *Cereal Chem.* **52,** 726.

Kleese, R. A., Rasmusson, D. C., and Smith, L. H. (1968). Genetic and environmental variation in mineral element accumulation in barley, wheat, and soybeans. *Crop Sci.* **8,** 591–593.

Koivistoinen, P., Nissinen, H., Varo, P., and Ahlstrom, A. (1974). Mineral element composition of cereal grains from different growing areas in Finland. *Acta Agric. Scand.* **24,** 327–333.

Lorenz, K., and Loewe, R. (1977). Mineral composition of U.S. and Canadian wheat and wheat blends. *J. Agric. Food Chem.* **25,** 806.

Lorenz, K., Loewe, R., Weadon, D., and Wolf, W. (1980). Natural levels of nutrients in commercially milled wheat flours. III. Mineral analysis. *Cereal Chem.* **57,** 65.

McCance, R. A., and Widdowson, E. M. (1942). Mineral metabolism on white and brown bread dietaries. *J. Physiol. (London)* **101,** 44.

Martin, H. F., and Halton, P. (1964). Addition of different iron salts to flour: The factor of rancidity. *J. Sci. Food Agric.* **15,** 464.

Miller, J. (1977). Study of experimental conditions for most reliable estimates of relative biological value of iron in bread. *J. Agric. Food Chem.* **25,** 154.

Murphy, G. M., and Law, D. P. (1974). Some mineral levels in Australian wheat. *Aust. J. Exp. Agric. Anim. Husb.* **14,** 663.

Nahapetian, A., and Bassiri, A. (1976). Variations in concentrations and interrelationships of phytate, phosphorus, magnesium, calcium, zinc, and iron in wheat varieties during two years. *J. Agric. Food Chem.* **24,** 947–950.

Oparin, I. A., and Oparina, N. V. (1969). Iron in vegetable food products of the Dagestan USSR. *Vopr. Pitan.* **28**(1), 79.

Peterson, D. J., Johnson, V. A., and Mattern, P. J. (1983). Evaluation of variation in mineral element concentrations in wheat flour and bran. *Cereal Chem.* **60,** 450.

Pomeranz, Y., and Dikeman, E. (1982). Mineral and protein content in hard red winter wheat. *Cereal Chem.* **59,** 139.

Ponte, J. G. (1979). Kansas State University studies on expanded fortification. *64th Annu. AACC Meet.* Abstract 104.

Ranum, P. M., and Loewe, R. J. (1978). Iron enrichment of cereals. *Baker's Dig.* **52**(3), 14.

Ranum, P. M., Barrett, F. F., Loewe, R. J., and Kulp, K. (1980). Nutrient levels in internationally milled wheat flours. *Cereal Chem.* **57**(5), 361.

Rao, S. K. (1974). Malnutrition in the eastern Mediteranean region. *W.H.O. Chron.* **28,** 172.

Rasmusson, D. C., Hester, A. J., Fick, G. N., and Byrne, I. (1971). Breeding for mineral content in wheat and barley. *Crop Sci.* **11,** 623–626.

Sayers, M. H., Lynch, S. R., Charlton, R. W., Bothwell, T. H., Walker, R. B., and Mayet, F. (1973). The effect of ascorbic acid supplementation on the absorption of iron in maize, wheat and soya. *Br. J. Haematol.* **24,** 209.

Schulerud, A. (1974). Iron enrichment of wheat flour. *Muehle + Mischfuttertech.* **11,** 145.

Shah, B. G., and Delonje, B. (1973). Bioavailability of reduced iron. *Nutr. Rep. Int.* **7,** 151.

Tweet, A. (1979). The production of traditional breads–Iran, Cuba, Syria, Japan. *Proc. Am. Soc. Bakery Eng. 55th Annu. Mtg.*, p. 38.

Waddel, J. (1973). "The Bioavailability of Iron Sources and their Utilization in Food Enrichment." FASEB Report for the FDA, Bethesda, Maryland.

Watson, C. A., Shuey, W. C., and Lacher, C. P. Mineral content of wheats and their milled products. Unpublished data.

Widhe, T. (1970). (Title unavailable.) *Liusmetal Steknic* **12,** 86.

World Health Organization. (1971). Joint FAO/WHO Expert Committee on Nutrition, 8th report, Food fortification, Protein calorie malnutrition. *W.H.O. Tech. Rep. Ser.*

Wyse, B. W., Sorenson, A. W., Wittner, A. J., and Hansen, R. G. (1976). Nutritional quality index identifies consumer nutrient needs. *Food Technol.* **30,** 22.

Yadav, S. P., Lal, B. M., and Gupta, Y. P. (1973). Chemical composition and protein quality of improved varieties of Indian wheat. *Indian J. Nutr. Diet.* **10,** 178.

Ziegler, E., and Greer, E. N. (1971). Principles of milling. *In* "Wheat: Chemistry and Technology" (Y. Pomeranz, ed.), p. 115. Am. Assoc. Cereal Chem., St. Paul, Minnesota.

7

Breakfast Cereals and Dry Milled Corn Products

RAY H. ANDERSON
General Mills, Inc.
James Ford Bell Technical Center
Minneapolis, Minnesota

I. INTRODUCTION

The addition of iron to ready-to-eat (RTE) breakfast cereals presents functional problems, particularly in the effects of added iron compounds on the shelf life of cereals. To clarify that effect, it is necessary first to explain what is meant by shelf life and the mechanism of oxidation that brings about deterioration in the quality of cereals after manufacture.

II. READY-TO-EAT CEREALS: OXIDATIVE RANCIDITY AND STALING

Commercial RTE breakfast cereals are adequately but not indefinitely shelf stable. Under normal commercial storage conditions they will be palatable during the period required for sale and consumption. Ultimately RTE cereal will deteriorate in odor and flavor by becoming rancid or stale.

Oxidative rancidity is favored by low moisture content, and if the water activity in a cereal does not increase beyond a point corresponding to about 8%

ISBN 0-12-177060-5

moisture content, the cereal may become rancid if its age exceeds its normal storage life. The critical moisture content of about 8% is only an approximate figure, as many other factors (including ingredients, processing conditions, initial moisture content, as well as the rate of increase of moisture content of the cereal) influence that value. The resistance of cereals to oxidative rancidity can be increased by the use of commercial antioxidants as additives, but the major deterrent seems to be products of browning reactions generated during the processing of the cereals (Anderson *et al.*, 1963).

Staling is ill defined and the mechanism little known. In RTE cereals, the process of staling includes increases in products of oxidation of fats, but the products seem to differ from those formed in oxidative rancidity. Staling is characterized more by disagreeable flavors than by off-odors. Oxidative rancidity is usually accompanied by strong disagreeable odors. Staling also involves textural toughening, whereas oxidative rancidity may have no effect on texture.

Staling has been postulated to be initiated by processes similar to oxidative rancidity, but in the presence of a relatively high water activity, products formed from the deterioration of fats are of a less volatile type than those generated by oxidative rancidity (Anderson *et al.*, 1963). Staling is inhibited by packaging cereals in materials that retard or prevent increases in moisture content of cereals during normal storage. Staling may also be retarded by adjusting the initial moisture content of a cereal to a lower level, but if taken too far the low moisture content will enhance the rate of oxidative rancidity.

The occurrence or rate of oxidative rancidity or staling in a cereal is not influenced solely by the amount of fat present (Anderson *et al.*, 1963), as some corn-based cereals of less than 1% ether-extractable fat will become rancid or stale just as quickly as some oat-based cereals of 7% ether-extractable fat under similar conditions.

Oxidative rancidity exhibits an induction period in a cereal during which time the indicators and products of rancidity increase only slowly, presumably caused by inhibition of the oxidative reactions by antioxidants normally present, added, or generated in processing (Lundberg, 1962b). After the induction period, the formation of products of oxidative rancidity increases at an accelerated rate. A free-radical mechanism is postulated for oxidative rancidity, as the extent of oxidative reaction, if plotted against time, indicates autoxidation that results from products of initiation or propagation of reactions themselves becoming initiators of further reactions.

III. MECHANISM OF OXIDATIVE RANCIDITY IN FATS OF READY TO EAT CEREALS

The "hydroperoxide" mechanism is generally accepted as the route by which fats autoxidize in a chain reaction schematically postulated as follows:

$$\text{RH (fatty acid)} \xrightarrow[\text{activator}]{-H^+} \text{R}\cdot \text{ (free radical)}$$

$$\text{R}\cdot + O_2 \longrightarrow RO_2^{\cdot} \text{ (peroxide radical)}$$

$$RO_2^{\cdot} + \text{R'H} \longrightarrow \text{R}'^{\cdot} + RO_2H \text{ (hydroperoxide)}$$

The hydroperoxide then breaks down to a number of oxidation products, and the chain reaction is further propagated by $R'^{\cdot}$ or other radicals (Swern, 1961).

The removal of hydrogen atoms from the chains of unsaturated fatty acids is necessary to initiate autoxidation. The direct reaction of oxygen with a fatty acid, as schematically portrayed in the following equation, does not take place, as it is an endothermic reaction requiring a high activation energy (Uri, 1961a).

$$\text{RH} + O_2 \rightarrow \text{R}\cdot + HO_2^{\cdot}$$

This reaction can occur only by means of energization of the hydrogen atom or oxygen. A number of activators will provide this energy, including ultraviolet light, heat, or heavy metals. Iron is an effective catalyst to initiate autoxidation. An iron content of 1 ppm in an unsaturated fat will initiate autoxidation at room temperature (Lundberg, 1962a).

IV. ACTION OF IRON IN OXIDATIVE RANCIDITY OF READY-TO-EAT CEREALS

Iron forms complexes and free radicals with oxygen that are energized sufficiently to remove hydrogen from the fatty-acid chains and form free-radical derivatives of fatty acids. The following equation simplistically portrays complex reactions of iron with oxygen that result in energized complexes and free radicals (Uri, 1961b):

$$Fe^{2+} + O_2 \rightarrow Fe^{3+} + O_2^{\dot{-}} \text{(superoxide)}$$

Haber–Weiss mechanism:

$$2\ O_2^{\dot{-}} + 2\ H^+ \rightarrow H_2O_2 + O_2$$

$$Fe^{2+} + H_2O_2 \rightarrow HO\cdot \text{ (hydroxyl radical)} + OH^- + Fe^{3+}$$

It has been shown with model physiological systems *in vivo,* as well as with nonphysiological model systems, that oxidation of lipids is greatly facilitated by the presence of iron whether the lipid peroxidation is dependent on the superoxide radical (Tien *et al.*, 1981a) or on the hydroxyl radical (Tien *et al.*, 1982). It is also evident in these reports that iron acts as an initiator of lipid peroxidation in different forms, depending on the system. Either ferrous or ferric iron or the chelation by EDTA of iron (Tien *et al.*, 1982) may initiate the peroxidation reactions, or the ADP–Fe^{2+}–O_2 (perferryl) complex (Tien *et al.*, 1981b) may be the effective initiator.

The presence of soluble reactive iron and oxygen catalyzes oxidative reactions of unsaturated fats in food products, including cereals, because of the formation of hydroxyl and superoxide free radicals during processing or storage of the products.

V. ENRICHMENT OF CEREAL GRAIN PRODUCTS WITH IRON

In 1941 the National Academy of Sciences recommended that certain products of the milling of grains be enriched with three vitamins and iron because there seemed to be a widespread need in the diets of U.S. people for these nutrients, and they were partially removed during milling and refining processes from some grain products. A review by Ranum and Loewe (1978) outlines and discusses the history and extent of and problems encountered in the enrichment of cereal grain flours with iron.

Flours from wheat or corn are quite stable to oxidative rancidity at ordinary commercial moisture contents, but are more prone to rancidity if dried for use in mixes to lengthen the shelf life of the leavening in the mixes. It was found that the addition of soluble reactive salts to grain flours, particularly to dried flours in mixes, tended to reduce substantially the shelf lives of the products because of oxidative rancidity. Either ferrous or ferric iron can initiate autoxidation of fats in the presence of oxygen (Uri, 1961b).

Elemental iron, reduced to that state from iron salts by action of hydrogen or carbon monoxide, was generally used to enrich the grain products with iron. Ferric orthophosphate or ferric sodium pyrophosphate could also be used without causing rancidity in grain flours. Ferrous sulfate could be used if added to grain flours before baking or cooking, but caused rancidity if added to flours at manufacture prior to normal shelf storage.

VI. RESTORATION OF READY-TO-EAT CEREALS WITH IRON

In 1941 manufacturers agreed to restore RTE breakfast cereals to whole-grain levels with the three vitamins and iron added to flours for enrichment. The cereals were restored as if the entire cereal were 100% whole grain. In some cereals this restoration was slight, because they were whole-grain cereals requiring restoration only for dilution with other ingredients or because of losses in processing.

It was found that, as in the case with flours from grains, cereals became rancid relatively quickly if soluble reactive iron salts such as ferrous sulfate were added to them during processing. Elemental iron, ferric sodium pyrophosphate, and ferric orthophosphate were satisfactory as iron additives. Elemental iron has several disadvantages in cereals. First, it causes a slight grayish color that can be partially ameliorated by slightly increased toasting. Second, it can gather on magnets that are in the cereal processing lines to remove tramp metal. Rearrangement of the magnets has minimized this problem. The disadvantages of elemental iron are outweighed by its inertness (in lack of reaction with oxygen), and consequently it is not as prone to cause oxidation of cereals as are soluble reactive sources of iron.

In 1955 the first cereal was introduced that was fortified with vitamins and iron beyond whole-grain restoration levels. Elemental iron was satisfactory for fortification of the cereal.

In 1961 a cereal was introduced that was fortified with vitamins and iron at 100% of the minimum daily requirement (MDR) per ounce of cereal of iron (10 mg) and of the vitamins that had a MDR, plus added vitamins B_6 and B_{12}. With high levels of added vitamins it became more necessary to consider the potential of iron, added in larger amounts, to affect the stability of added vitamins. Elemental iron or ferric phosphates did not seem to cause deterioration of vitamins.

There was some loss of sensitive vitamins in stored cereals to which reactive iron salts had been added (R. H. Anderson, unpublished data). These losses could be ascribed to actions of free radicals formed during autoxidation of fats and not necessarily to any other action of the iron itself other than as a catalyst for autoxidation of cereal fats.

During research on the effects of different forms of iron added to cereals, a number of commercially available forms on the GRAS (generally recognized as safe) list were tested (R. H. Anderson, unpublished data), including ferric ammonium citrate, ferrous fumarate, ferric orthophosphate, ferric sodium pyrophosphate, ferric choline citrate, and ferrous sulfate. It became apparent that the extent to which the ferrous or ferric ion was complexed to prevent its free radical-forming reactions with fats was directly related to stability of the product fortified with iron. Cereals fortified with ferrous sulfate became rancid during normal shelf storage. Those fortified with elemental iron were sufficiently stable. In between, largely satisfactory but more expensive to varying degrees than elemental iron, were the complexes of iron named above. Products made with ferric sodium pyrophosphate were as stable to rancidity as those containing added elemental iron. Although more expensive than elemental iron, ferric sodium pyrophosphate did not cause discoloration or problems with magnets, so it was used to a large extent in RTE cereals.

In about 1971, most cereals were fortified with vitamins and iron at levels of 33% of MDR. Later when the U.S. Recommended Daily Allowances (USRDAs) wre established, fortification of most cereals was changed to 25% of USRDAs of vitamins and iron. For these levels of fortification and for those up to 100% of USRDA, elemental iron, sodium ferric pyrophosphate, or ferric orthophosphate, providing up to 18 mg of iron per ounce of cereal, was satisfactory for use. Some published human experiments on the biological availability of various forms of iron added to foods were indicating in the 1970s that the iron of ferric orthophosphate and ferric sodium pyrophosphate was not as biologically available as that of elemental iron in baked goods from wheat flour containing added iron. For that reason the cereals that had been fortified with ferric sodium pyrophosphate were changed over to fortification with elemental iron. The change from ferric sodium pyrophosphate to elemental iron was voluntary. No regulations existed in the United States that stipulated the form of iron to be used in fortification of foods.

The studies that led to that change of form of iron fortification are discussed in detail in Chapters 1 and 2 of this book, so they are not discussed or cited here.

Elemental iron or iron salts may be added to RTE cereals by mixing the source of iron with the cereal ingredients before cooking. In some cereal systems, the iron can be added to the cooked cereal dough or mash before extrusion and cutting of the cooked dough into pellets. In other cereal systems there is no step between cooking and extrusion, so the iron must be added with the ingredients before cooking.

A flaked RTE cereal system and one of a puffed cereal are schematically diagrammed here to show crudely the steps involved in processing to illustrate opportunities, or lack of opportunity to add iron during various stages of processing.

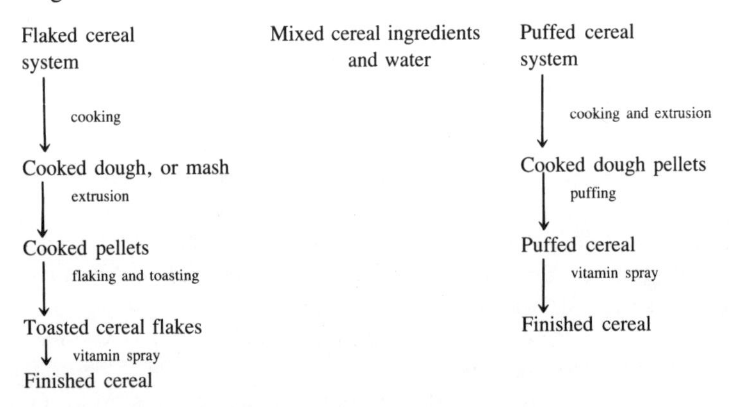

VII. USE OF ELEMENTAL IRON TO FORTIFY READY-TO-EAT CEREALS

At the present time, most RTE cereals are fortified with hydrogen-reduced iron powder or with electrolytically precipitated iron powder.

There are three forms of elemental iron powder available for iron fortification of food products. These are hydrogen-reduced, electrolytically precipitated, and carbonyl iron powders. Of the three, carbonyl iron is the most expensive and is not used in cereals at present. Some rat and human studies of iron availability have indicated that electrolytically precipitated iron powder and carbonyl iron may be more biologically available in certain baked products than hydrogen-reduced iron powder by virtue of a finer particle size.

VIII. RELATIVE BIOLOGICAL VALUE OF IRON ADDED TO READY-TO-EAT CEREALS

There are no published data from rat or human studies on the relative biological value (RBV) of different sources of iron in RTE cereals. Since they are cereal grain-based products, it might be assumed that the RBV of the different sources in cereals is similar to that in bread, but such an assumption cannot be readily made, as Wood *et al.* (1978) found that cooking processes combining heat and pressure increased the RBV of ferric phosphates tested. Processing of cereals includes cooking of cereal doughs at relatively high temperatures and pressure (15 psi or higher) compared to baking of bread. These processes also include toasting processes in air ovens in the case of flaked cereals, or under pressure (60 psi or higher) in puffing guns for puffed cereals. The processes of RTE cereal preparation are very different from baking processes for bread and biscuits, and the RBV of sources of iron in RTE cereals may not be similar to RBVs of the same sources in bread. Furthermore, Lee and Clydesdale (1979, 1980) have shown that the state of iron in a finished processed food may be very different from the added form. In biscuits or bread the iron from either added reduced iron powder or ferrous sulfate was mostly insoluble or in complex forms. Very little from either source was in the soluble ionic form. The final forms of the iron from six GRAS sources of iron baked in biscuits or bread were very similar in being almost all insoluble or complexed forms. Added iron in different forms is changed considerably in state during processing, and it may be difficult to interpolate the RBV of a particular source of iron from one form of food processing to another.

IX. EFFECTS OF INSOLUBLE OR COMPLEXED IRON IN READY-TO-EAT CEREALS

Although in the reports of Lee and Clydesdale (1979, 1980) the final form of iron from six different sources, including ferrous sulfate, was largely insoluble and complexed with little or no soluble ionic forms, the chemical nature of the insoluble and/or complexed forms resulting from processing of different sources in biscuits or bread might be different.

If it is assumed that a similar phenomenon would be encountered in cereals, the relative stability to rancidity of cereals containing different sources of iron processed in the cereals may be due to one of two reasons. If final insoluble and/or complexed forms of iron from different sources in cereals are similar in chemical activity after processing, instability due to some soluble ionic sources may be due to precursors formed in fats during processing that can initiate autoxidation of the fats during storage of the cereals. Alternatively, sources of iron like ferrous sulfate, originally soluble and reactive, are converted to insoluble and/or complexed forms that are more prone to form free radicals and initiate oxidative rancidity than are complexes from nonreactive or sequestered iron in fats of cereals.

X. FUNCTIONAL EFFECTS VERSUS RELATIVE BIOLOGICAL VALUE OF IRON IN READY-TO-EAT CEREALS

For RTE cereals, elemental iron powder, and secondarily ferric sodium pyrophosphate and ferric orthophosphate, are the most practical sources of iron from considerations of cost, shelf stability, and quality of products. Consideration of use of any other source of iron based on RBV in cereals should consider effects of the iron source on acceptability of cereals as well as the RBV of the incorporated iron.

XI. DRY MILLED CORN IRON ENRICHMENT

Since cereal enrichment standards were applied to wheat flour and bread in 1941, enrichment standards have also been set for corn and rice. For iron the range of enrichment is 13–26 mg iron/lb bolted meal, degermed meal, or grits. Contents of iron naturally present in these products in milligrams per pound of product are 8.2 in bolted meal, 5.0 in degermed corn, and 4.5 in grits (Ranum and Loewe, 1978). Whole corn contains 10.9 mg iron/lb and does not have an enrichment standard.

Cornmeal and grits of commerce are fairly similar to wheat flour in moisture content and are normally sufficiently shelf stable but prone to rancidity if soluble reactive iron salts such as ferrous sulfate are added for enrichment. Little work has been published on effects of various sources of iron on the shelf life of cornmeals and grits. Information obtained in extensive studies on wheat flours has been applied to the enrichment of corn products with iron. At the present time about three-fourths of the corn meals and grits of commerce are enriched with elemental iron, mostly iron powder reduced by the action of hydrogen or carbon monoxide in iron salts (Ranum and Loewe, 1978). Elemental iron powder does not cause difficult functional problems of separation, color, or rancidity in cornmeals or grits.

Ferric phosphate or ferric sodium pyrophosphate could be used instead of iron without inducing rancidity or causing other functional problems in cornmeals and grits, but use of these sources of iron has been largely discontinued in cereal products since human experiments in the 1970s indicated that these ferric phosphates have a lower RBV than elemental iron.

Bell (1974) prepared and patented a stabilized ferrous sulfate called Bio-Iron, which was tested for use in the enrichment of wheat flour and corn products. Anderson *et al.* (1976) added reduced iron or stabilized ferrous sulface (Bio-Iron) to cornmeal, grits, and soy-fortified cornmeal at levels of 40 and 200 mg iron/lb of corn product. After 56 days at 49°C, the meal and grits fortified with Bio-Iron showed development of rancidity; those products containing reduced iron powder did not. Soy-fortified meal containing either source of iron was stable under the conditions of storage.

Little work has been published on the RBV of iron sources of enrichment in corn products. A great deal of work has been done on wheat products, and it has been assumed that the RBV of various forms of iron enrichment in corn products would be relatively similar to those values in wheat products.

Miller (1976) found that, when fed to rats, iron naturally present in uncooked whole corn has a RBV of half that of reference ferrous sulfate. Whether or not the RBV of the naturally occurring iron is diminsihed by complexing with the phytates in corn is not known. A number of studies of possible complexing of phytates with iron in wheat products have been cited by Ranum and Loewe (1978), and the issue is in doubt whether or not the presence of phytates has an effect in diminishing the RBV of the natural iron present or of added iron.

It cannot be assumed without some reservations that the RBV of sources of iron enrichment in processed products made from cornmeals or grits would be relatively similar to those values in wheat flour.

Lee and Clydesdale (1979, 1980) have shown that the forms of iron in finished processed cereal products can differ widely from original sources of iron. If processes for products differ in temperature, pressure, or the presence of ingredients that might react or complex with iron, the state of iron in those products can be different, although derived from the same source of iron added to each product.

Although it is not known by experimentation what the RBV of many sources of iron enrichment might be in processed products from meals or grits, it seems reasonable to assign relative ratings corresponding to those in processed products from wheat fllours. Processes of products from wheat flours and from cornmeals are similar enough to warrant a rough degree of interpolation of RBVs of added iron sources. It seems reasonable, based on present information and considering a balance of functional practicality with RBV, that a form of elemental iron is the preferred source of iron for enrichment of cornmeals and grits.

REFERENCES

Anderson, R. A., Vojnovitch, C., and Bookwalter, G. N. (1976). Iron enrichment of dry-milled corn products. *Cereal Chem.* **53,** 937–946.

Anderson, R. H., Moran, D. H., Huntley, T. H., and Holahan, J. L. (1963). Responses of cereals to antioxidants. *Food Technol.* **17,** 115–120.

Bell, A. D. (1974). U.S. Patent 3,803,292.

Lee, K., and Clydesdale, F. M. (1979). Iron sources used in food fortification and their changes due to food processing. *CRC Crit. Rev. Food Sci. Nutr.* **11,** 117–153.

Lee, K., and Clydesdale, F. M. (1980). Effect of baking on the forms of iron in iron-enriched flour. *J. Food Sci.* **45,** 1500–1504.

Lundberg, W. O. (1962a). Oxidative rancidity and its prevention. *In* "Autoxidation and Antioxidants" (W. O. Lundberg, ed.), Vol. 2, p. 457. Wiley (Interscience), New York.

Lundberg, W. O. (1962b). Oxidative rancidity and its prevention. *In* "Autoxidation and Antioxidants" (W. O. Lundberg, ed.), Vol. 2, p. 464. Wiley (Interscience), New York.

Miller, J. (1976). Uncooked grain corn as a source of iron for normal and anemic rats. *Cereal Chem.* **53,** 413–420.

Ranum, P. M., and Loewe, R. J. (1978). Iron enrichment of cereals. *Baker's Dig.* **52,** 14–20.

Swern, D. (1961). Primary products of olefinic oxidation. *In* "Autoxidation and Antioxidants" (W. O. Lundberg, ed.), Vol. 1, p. 30. Wiley (Interscience), New York.

Tien, M., Svingen, B. A., and Aust, S. D. (1981a). Superoxide dependent lipid peroxidation. *Fed. Proc., Fed. Am. Soc. Exp. Biol.* **40**(2), 179–182.

Tien, M., Svingen, B. A., and Aust, S. D. (1981b). Initiation of lipid peroxidation by perferryl complex. *In* "Oxygen and Oxy-Radicals in Chemistry and Biology" (M. A. J. Rodgers and E. L. Powers, eds.). Academic Press, New York.

Tien, M., Svingen, B. A., and Aust, S. D. (1982). An investigation into the role of hydroxyl radical in xanthine oxidase-dependent lipid peroxidation. *Arch. Biochem Biophys.* **216**(1), 142–151.

Uri, N. (1961a). Physio-chemical aspects of autoxidation. *In* "Autoxidation and Antioxidants" (W. O. Lundberg, ed.), Vol. 1, pp. 77–86. Wiley (Interscience), New York.

Uri, N. (1981b). Physio-chemical aspects of autoxidation. *In* Autoxidation and Antioxidants" (W. O. Lundberg, ed.), Vol. 1, pp. 93–96. Wiley (Interscience), New York.

Wood, R. J., Stake, P. E., Eiseman, J. H., Shippee, R. L., Wolske, K. E., and Koehn, U. (1978). Effects of heat and pressure processing on the relative biological value of selected dietary supplemental inorganic iron salts as determined by chick hemoglobin repletion assay. *J. Nutr.* **108,** 1477–1484.

8

Iron Enrichment of Rice

JOHN W. HUNNELL
Research and Development
Riviana Foods, Inc.
Houston, Texas

K. YASUMATSU AND S. MORITAKA
Food Products Division
Takeda Chemical Industries, Inc.
Tokyo, Japan

I. INTRODUCTION

The cereal grains, including rice, represent the single largest source of calories in the world. Unfortunately, there has been a tendency for research to be aimed at providing appropriate fortification of those grains most widely used in the affluent countries. This has resulted in a paucity of knowledge concerning the technology of iron fortification in rice to achieve the greatest bioavailability along with optimal functional properties for food compatibility.

In the Western world, the addition of iron to rice is dependent in general on a premix supplied by vendors to the rice industry. In the developing world many attempts to fortify rice have been made with varying degrees of success. Because of the problems involved, there has been a tendency to fortify rice with vitamins and include an inert form of iron, if any. Hopefully, research will lead us to a better situation than this.

Rather than discuss every attempt to fortify rice in every country, the editors decided to provide information from two countries by experts in those countries,

In Section II of this chapter, Dr. Hunnell describes fortification procedures in

ISBN 0-12-177060-5

the United States for permitted nutrients. In Section III Drs. Yasumatsu and Moritaka describe a case history of the development in Japan of an enriched rice blend to be added to polished rice to improve its overall nutritional contribution. These contributors provide some interesting insights and ideas in iron–rice technology and offer a challenge to those working in the field worldwide.

II. UNITED STATES

The use of iron and other enrichment materials for rice in the United States is controlled by federal regulation. The regulations are optional, but if used, prescribe the levels that must be used and the information that must be declared in the package labeling information. Several states have made the optional federal regulations mandatory. Most of the rice sold today in the United States is enriched, because almost all interstate and national marketers of rice enrich their products to avoid the need for separate inventories.

The intent of the regulations is to return the milled rice to the nutritional level of brown rice (unmilled rice), and therefore—within the regulations—riboflavin is specified along with thiamine, niacin, and iron. However, because of the negative consumer response to the yellow color of riboflavin, it is optional and in general it is not used. Vitamin D and calcium are also optional. Furthermore, because of the nature of the parboiling process, parboiled rice does not require the addition of niacin to comply with the standard. Parboiled rice also may contain BHT to a prescribed maximum level of 0.0033% to serve as an antioxidant. Harmless carriers may be employed in quantities necessary to accomplish intimate and uniform mixtures of the enriching components and rice.

The standard of identity for rice requires that each pound of milled rice, if enriched, must contain not less than 2.0 and not more than 4.0 mg of thiamine, not less than 16 and not more than 32 mg of niacin or niacinamide, and not less than 13 and not more than 26 mg of iron (Fe). If riboflavin is used, it is limited to not less than 1.2 and not more than 2.4 mg. If vitamin D is used it must not be less than 250 and not more than 1000 USP units. Similarly, if calcium (Ca) is used it must be at levels not less than 500 and not more than 1000 mg.

The regulation requires that iron and calcium may be added only in forms that are harmless and assimilable. Because of interaction with components of rice and the resultant poorer quality, the preferred form of iron for rice enrichment is ferric orthophosphate. Reduced iron has been used but is less preferred because of its sensitivity to magnets employed in processing for the removal of tramp metal. Other more available forms are not functionally suitable.

Enrichment is accomplished with rice in the United States in one of two ways. An enrichment blend may be mixed with rice as a powder and, depending on the commercial mixture used, added at a rate of 1 lb to 3000 or 6000 lb of rice. It is readily apparent that uniformity can be a considerable problem. A better approach is to use a coated broken-grain product protected with a coating such as ethyl cellulose, a so-called premix. The broken grains can be precooked and used as a substrate for the enrichment of precooked rice as well as uncooked brokens for conventional milled rice. This material is applied at a level of 0.5% or 0.5 lb in 99.5 lb of rice.

Attempts to market a rice enrichment product made by coextruding rice flour and the enrichment compounds have not been successful. However, the concept based on prototypes appears to be feasible.

In the United States, the standard requires that the label include the statement: "To retain vitamins, do not rinse before or drain after cooking." Alternatively, the packer may choose not to use this statement and to demonstrate that no less than 85% of the minimum quantities of the enriching substances remains after cooking by a prescribed method. "Enriched rice" or "enriched parboiled rice" must become part of the name of the product and must be shown on the principal display panel of the package. Furthermore, the use of enrichment in rice requires the use of the nutritional label format on the package. Brown rice and wild rice are not enriched.

Technical problems remain. These can be solved by additional research on iron–rice interactions and the development of functionally suitable and bioavailable iron fortificants to produce uniform and stable enriched rice. However, almost no attention is being given to this aspect of iron fortification within industry at the present time.

III. INTERNATIONAL CONSIDERATIONS

In Japan, the consumption of rice has shown a steady decrease. Yet the Japanese depend on rice for 37% of their energy and 17% of their protein supply (Ministry of Health and Welfare, 1982). This means that, in Japan, rice remains as the staple food. The Japanese prefer it in its polished state, whereby essential nutrients such as vitamin B_1 may be lost. Therefore, vitamin B_1-enriched rice has been used to blend with polished rice at a ratio of 1:200 for over 30 years, contributing to the improvement of the nutritional state of the Japanese people.

However, anemia has also been recognized as a significant public health problem in Japan. According to the results of the National Survey on Nutrition (Ministry of Health and Welfare, 1980), about 20% of adult women are anemic. Most of this anemia is considered to be due to iron deficiency (Uchida, 1975). Therefore, iron enrichment of rice in Japan is an important consideration, even though only a few studies have been reported. Furter *et al.* (1946) and Mickus

(1955) reported on processing methods for the fortification of rice with vitamin B_1, niacin, and iron, but they did not report on the clinical evaluation of such rice on iron deficiency. Peil *et al.* (1981) also reported on a processing method for the production of fortified rice suitable for cooking in excess water and draining. The fortified rice contained vitamins A, B_1, and B_2, niacin, and iron, but, again, its clinical effect was not reported.

The authors have developed new types of enriched rice that contain vitamins B_1, B_2, and B_6, niacin, pantothenic acid, vitamin E, and minerals such as calcium and iron. This enriched rice is blended with polished rice at a ratio of 1:200, achieving near-equality with eight nutrients in brown rice. The new enriched rice, *Shingen,* has been marketed since April 1981 in Japan. *Shingen* means "brown rice in the new age." The processing method, its properties, and the clinical effects of the new enriched rice are described herein.

A. The New Enriched Rice, *Shingen*

Many of the micronutrients in brown rice are contained in the germ and the bran, which are removed in the manufacture of the polished rice eaten as a staple food in Japan.

According to the results of a consumer survey carried out by our company, brown rice is perceived as nutritious, natural, and healthful, but hard to cook, hard to eat, and unpalatable. The new enriched rice *Shingen* was developed with a view to optimizing the advantages of brown rice while overcoming some of the disadvantages such as being hard to cook, hard to eat, and unpalatable.

B. Processing Methods

In Japan, eight iron compounds are permitted for use in foods (Table I). However, only ferric pyrophosphate does not affect the appearance, aroma, and flavor of cooked rice. Therefore, it is often chosen as the source of iron for enriched rice. Furter *et al.* (1946) and Mickus (1955) also used this iron compound in the studies mentioned previously. Peil *et al.* (1981) used electrolytically reduced iron because of its greater bioavailability. Their fortified rice is designed to retain added nutrients without loss during cooking in excess water. They reported that iron retention was 100% when 1 g of the fortified rice was cooked in 100 ml water and drained. It is assumed that reduced iron does not impair the flavor of cooked rice, because it is completely retained in the fortified rice during cooking in an inert form.

Dibenzoyl thiamine hydrochloride (DBT) is used for enriched rice as well as for the vitamin B_1-enriched rice *Poly Rice* which has been marketed for over 30 years in Japan. Almost colorless and tasteless, DBT is barely soluble in water (below 50°C), but it is soluble in hot water or acidic water.

Table I
IRON COMPOUNDS PERMITTED FOR USE IN FOOD IN JAPAN[a]

Ferrous sulfate
Ferrous lactate
Ferrous pyrophosphate
Ferric chloride
Ferric citrate
Ferric ammonium citrate
Sodium ferric succinic citrate
Ferric pyrophosphate

[a]From Ministry of Health and Welfare (1978).

In Japan, *dl*-α-tocopherol is permitted for use only as an antioxidant in oils and foods (Ministry of Health and Welfare, 1978). Therefore, natural vitamin E extracted from soybeans and concentrated to such an extent that α-tocopherol is greater than 50% of the total tocopherol, is used.

Food additive grade riboflavin, nicotinamide, pyridoxine hydrochloride, calcium pantothenate, and calcium carbonate are also used.

Typical manufacturing methods are used for our enriched rice, such as the acid-parboiling rice method and the coating method. The former was developed by the Laboratory of Nutritional Chemistry, Faculty of Agriculture, Kyoto University (Kondo *et al.*, 1949a,b, 1950, 1951). This laboratory had been studying manufacturing methods for enriched rice that would suit the taste of Japanese people. Their method established the basis for manufacturing vitamin B_1-enriched rice in Japan. The polished rice is soaked in an acidic solution containing the required amounts of vitamins B_1 and B_2, at a temperature of about 40°C. The soaked rice is then steamed for a very short time in superheated steam, and dried in an air draft with filtered clean air, making adjustments so as to obtain a finished product with a moisture content of less than 14%.

The coating method was developed originally by Williams and co-workers, who synthesized thiamine for the first time, and was commercialized by Hoffman-La Roche (Furer *et al.*, 1946; Salcedo *et al.*, 1950). In this method, the vitamin solution is sprayed over the rice grains, then a protective coating, which does not dissolve in cold water but melts in hot water above 70°C, is applied to prevent loss of vitamins through rinsing.

Since our newly developed enriched rice contains vitamin E, calcium, and iron, all of which are almost insoluble in water, it cannot be manufactured using only the acid-parboiled rice method. In Japan rice is cooked in water, which is completely absorbed. Therefore, when the enriched rice is prepared using only

Table II
AVERAGE NUTRIENT COMPOSITION OF THE ENRICHED RICE, *SHINGEN* (mg/g)

Vitamin B_1	1.5
Vitamin B_2	0.06
Niacin	6.2
Pantothenic acid	2.34
Vitamin B_6	0.08
Vitamin E	1.38
Calcium	8.0
Iron	1.2

the coating method, it is not gelatinized satisfactorily through cooking and differs from ordinary cooked rice in its hardness, Therefore, a two-step method was developed. First, an intermediate that contains vitamins B_1, B_2, and B_6, niacin, and pantothenic acid, is manufactured with the acid-parboiled rice method. It is then coated with vitamin E, calcium, iron, and a protective coating material. Adopting this method, the rice grains of the enriched rice are gelatinized com-

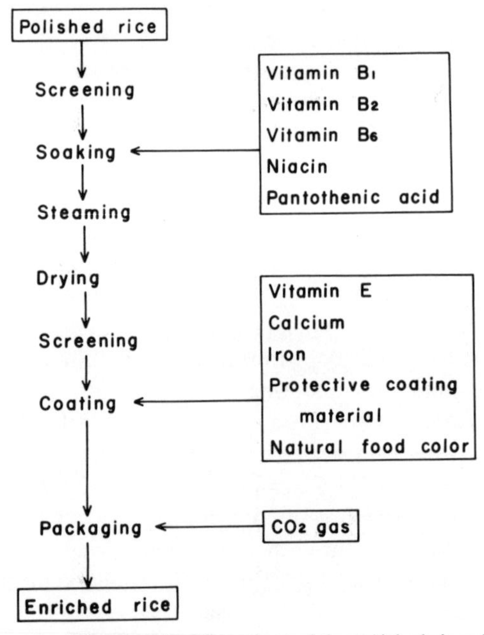

Figure 1. Manufacturing flow sheet of the enriched rice, *Shingen*.

Table III

AVERAGE NUTRIENT COMPOSITION OF RICE (mg/100 g)

Nutrient	Polished rice[a]	Brown rice[a]	Blended rice[b]
Vitamin B_1	0.12	0.54	0.87
Vitamin B_2	0.03	0.06	0.06
Niacin	1.40	4.50	4.50
Pantothenic acid	0.23	1.40	1.40
Vitamin B_6	0.06	0.43	0.10
Vitamin E	0.09	0.78	0.78
Calcium	6.0	10.0	10.0
Iron	0.5	1.1	1.1

[a]From Fukuba (1978), Iwai and Yokomizo (1967), Resource Council of Science and Technology Agency (1980), and Yasumoto *et al.* (1976).

[b]Mixture of 200 parts of the polished rice and 1 part of the enriched rice, *Shingen.*

pletely through cooking, and their texture is the same as that of ordinary cooked rice.

Very sophisticated coating techniques may be used in the manufacture of the enriched rice. For example, the stability of vitamin E may be improved by a technique in which vitamin E and iron are located in separate layers. Such a coating material does not dissolve in cold water but melts completely in hot water

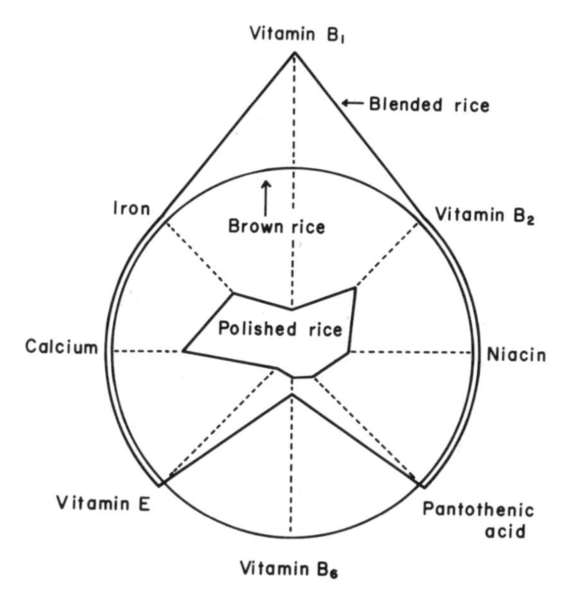

Figure 2. Comparison of the nutrient contents in polished, brown, and blended rice. Blended rice is a mixture of 200 parts of polished rice and 1 part of enriched rice, *Shingen.*

Table IV
RESULTS OF CONSUMER SURVEY ON THE ENRICHED RICE

	Consumer response (%)	
Parameter	Almost the same as ordinary cooked rice	Somewhat inferior to ordinary cooked rice
Appearance	95.5	4.5
Aroma	96.4	3.6
Taste	95.5	4.5

at a temperature above 70°C and therefore prevents the loss of vitamins and minerals during washing before cooking. This protective coating material is different from that used by Furter *et al.* (1946) and Mickus (1955) in their enriched rice.

The average nutrient composition and the manufacturing flow sheet for the new enriched rice *Shingen* are shown in Table II and Fig. 1, respectively.

C. Properties of the Enriched Rice

1. NUTRITIVE RESTORATION

The final quantities of eight nutrients in the blended rice, which is a mixture of 200 parts of the polished rice and 1 part of enriched rice, are nearly equal to those of brown rice (Table III, Fig. 2). Because 1 g of the enriched rice contains 1.0–1.5 mg of vitamin B_1, the blended rice is higher in this nutrient than is brown rice. On the other hand, the vitamin B_6 content of the blended rice is lower than that of brown rice, because of the undesirable effects of vitamin B_6 (pyridoxine hydrochloride) on the flavor of cooked rice. However, the amounts of all eight nutrients in the blended rice are higher than those in polished rice.

Table V
LOSS OF VITAMINS ON COOKING[a]

Vitamin	Remaining ratio (%)
Vitamin B_1	89
Vitamin B_2	88
Niacin	92
Pantothenic acid	97
Vitamin B_6	100
Vitamin E	85

[a]Measured at Japan Food Research Laboratories.

2. INFLUENCE ON THE FLAVOR OF COOKED RICE

Our Food Research Laboratories have carried out many studies on the flavor improvement of rice, the mechanism of storage deterioration, and a method for evaluating rice quality, because when the nutrition of polished rice is restored to that of brown rice, the flavor of the cooked rice must be changed. Each step in the manufacturing process was carefully examined for its effect on the flavor of cooked rice, and it was found that when the enriched rice is blended with polished rice in a ratio of 1:200, the flavor of cooked rice is unchanged.

Table IV shows the results of a consumer survey, in which enriched rice was delivered to 450 households in Fukuoka Prefecture. Nearly all respondents said that the addition of the enriched rice did not affect the appearance, aroma, or the taste of cooked rice,

3. LOSS OF NUTRIENTS THROUGH WASHING

As shown in Table V, the cooking loss of the vitamins contained in the enriched rice is about 10%, and storage stability is improved by the use of

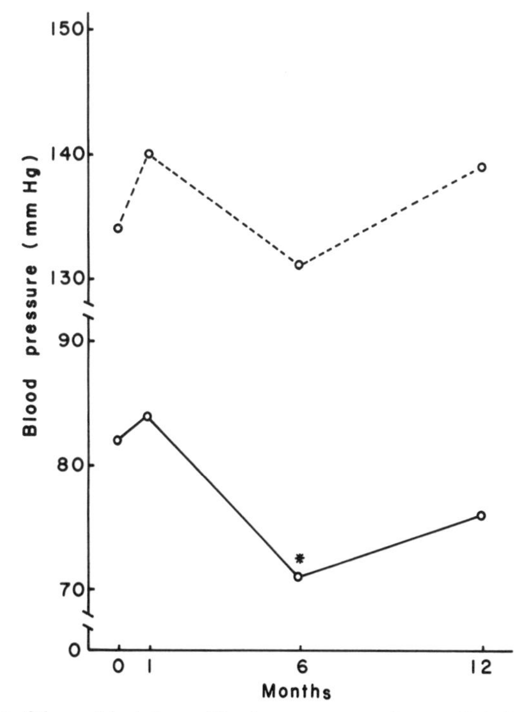

Figure 3. Effect of the enriched rice on blood pressure over a 1-year clinical study. *Significantly different from the initial value at 5% level. Solid line, diastolic blood pressure; dashed line, systolic blood pressure. [From Koyanagi *et al.* (1982).]

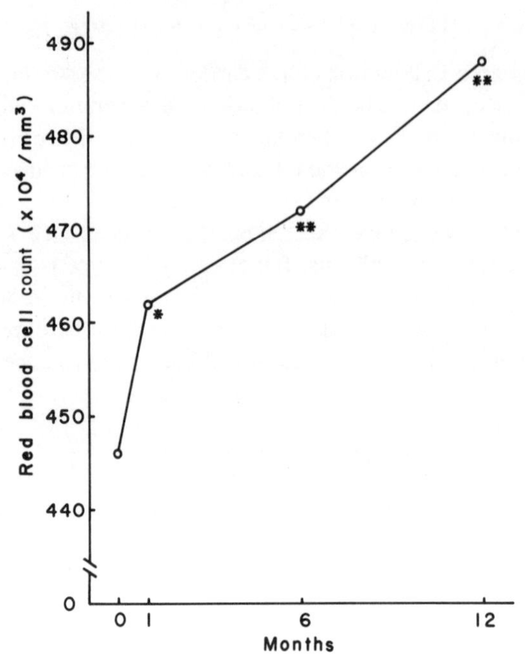

Figure 4. Effect of the enriched rice on red blood cell count over a 1-year clinical study. *Significantly different from the initial value at 5% level; **significantly different from the initial value at 1% level. [From Koyanagi *et al.* (1982).]

sophisticated coating techniques. Also, our enriched rice *Shingen* is packaged by a CO_2 gas exchange method (Mitsuda *et al.*, 1972) to improve the storage stability of vitamin E even further.

D. Clinical Studies with the Enriched Rice

Koyanagi *et al.* (1982) reported on a clinical study using the enriched rice. They chose 20 women from a farm village in Iwate Prefecture aged 49–74, who were served enriched rice for 1 year. Blood pressure (systolic and diastolic), red blood cell count, hemoglobin concentration, and the α-tocopherol content in the blood were examined, among other parameters.

The diastolic blood pressure showed a significant decrease after 6 months, whereas the systolic blood pressure was almost unchanged (Fig. 3). The red blood cell count increased significantly (from 445 to 489 $\times$ $10^4/mm^3$; Fig. 4). Hemoglobin also increased, from 13.3 to 14.8 g %, at the twelfth month (Fig. 5).

As shown in Fig. 6, the α-tocopherol level in the blood also increased significantly through the consumption period.

This case study, although not complete in all aspects, provided some interest-

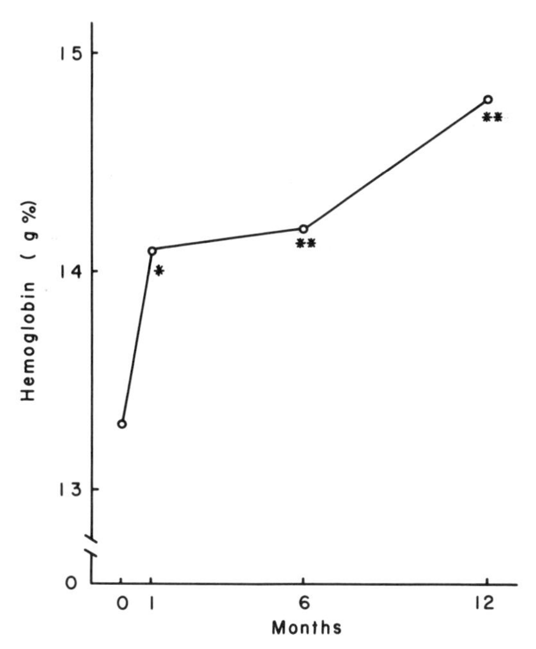

Figure 5. Effect of the enriched rice on hemoglobin levels over a 1-year clinical study. *Significantly different from the initial value at 5% level; **significantly different from the initial value at 1% level. [From Koyanagi *et al.* (1982).]

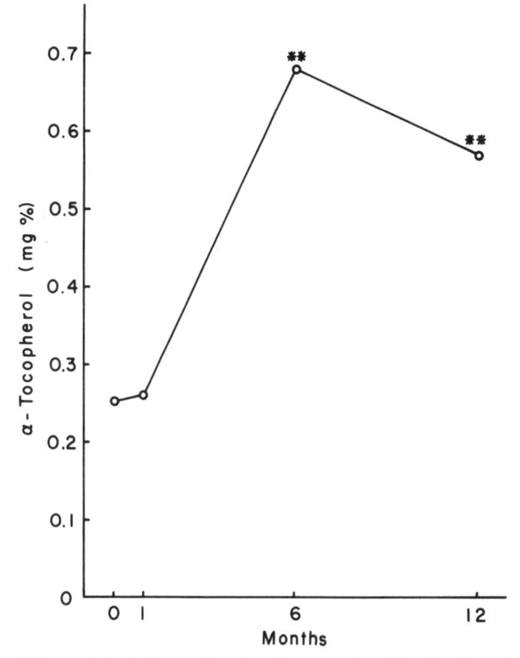

Figure 6. Effect of the enriched rice on levels of α-tocopherol in the blood over a 1-year clinical study. **Significantly different from the initial value at 1% level. [From Koyanagi *et al.* (1982).]

ing ideas in the technology of rice fortification that might be considered in combatting both iron deficiency and generally low nutrient levels in the diet.

REFERENCES

Fukuba, H. (1978). Haigamai. *Chomi Kagaku* **11,** 51–54.

Furter, M. F., Lauter, W. M., DeRitter, E., and Rubin, S. H. (1946). Enrichment of rice with synthetic vitamins and iron. *Ind, Eng. Chem.* **38**(5), 486–493.

Iwai, K., and Yokomizo, H. (1967). The distribution of B-vitamins in the rice grains. *J. Jpn. Soc. Food Nutr.* **20,** 495–499.

Kondo, K., Mitsuda, H., and Iwai, K. (1949a). Studies on so-called acid parboiled rice, *Vitamins (Japan)* **2,** 134–135,

Kondo, K., Mitsuda, H., and Iwai, K. (1949b). Studies on so-called acid parboiled rice (2). *Vitamins (Japan)* **2,** 264.

Kondo, K., Mitsuda, H., and Iwai, K. (1950). Studies on the enrichment of parboiled rice. *Vitamins (Japan)* **3,** 155.

Kondo, K., Mitsuda, H., and Iwai, K. (1951). Studies on the enrichment of white rice. *Vitamins (Japan)* **4,** 203–204.

Koyanagi, T., Chiba, S., Takanohashi, T., Oikawa, K., Akazawa, N., Tsunematsu, M., Kimura, T., and Koyama, K. (1982). A long-term feeding experiment of a new enriched rice in aged rural people. *36th Annu. Meet. Jpn. Soc. Food Nutr.*

Mickus, R. R. (1955). Seals enriching additives on white rice. *Food Eng.* **27**(11), 91–93.

Ministry of Health and Welfare, Tokyo (1978). "The Japanese Standards of Food Additives," 4th ed.

Ministry of Health and Welfare, Tokyo (1980). "Annual Report on the Nutritional Survey."

Ministry of Health and Welfare, Tokyo (1982). "Annual Report on the Nutritional Survey."

Mitsuda, H., Kawai, F., and Yamamoto, A. (1972). Underwater and underground storage of cereal grains. *Food Technol.* **26**(3), 50–56.

Peil, A., Barrett, F., Rha, C., and Langer, R. (1981). Retention of micronutrients by polymer coatings used to fortify rice. *J. Food Sci.* **47,** 260–262.

Resource Council of Science and Technology Agency, Tokyo (1980). "Standard Tables of Food Composition in Japan," Tentative revised ed.

Salcedo, J., Jr., Bamba, M. D., Carrasco, E. O., Chan, G. S., Conception, I., Jose, F. R., DeLeon, J. F., Oliveros, S. B., Pascual, C. R., Santiga, L. C., and Valenzuela, R. C. (1950). Artificial enrichment of white rice as a solution to endemic beriberi. *J. Nutr.* **42,** 501–523.

Uchida, A. (1975). Endemic and clinical studies on anemia in rural community. *Noson Igaku* **24,** 324.

Yasumoto, K., Iwami, K., Tsuji, H., Okada, J., and Mitsuda, H. (1976). Bound forms of vitamin B_6 in cereals and seeds. *Vitamins (Japan)* **50,** 327–333.

9

Fortification of Infant Formula

RICHARD C. THEUER
Research & Development
Beech-Nut Nutrition Corporation
Fort Washington, Pennsylvania

I. INTRODUCTION

Iron undernutrition is the most common nutritional problem of the American infant and toddler (Committee on Nutrition, American Academy of Pediatrics, 1969). This problem is primarily related to a high iron requirement: "During infancy (3 to 24 months) because of the rapid rate of growth, iron requirements in proportion to food intake exceed those of any other period of life" (Committee on Iron Deficiency, Council on Foods and Nutrition, American Medical Association, 1968).

Iron undernutrition is more likely among the poor (Owens *et al.*, 1971). Infants who are exclusively breast-fed rarely become anemic (Siimes *et al.*, 1984). The infant who is not breast-fed normally is fed an infant formula as the sole source of nutrition for 4–6 months; physicians rely on iron-fortified formula to provide bioavailable iron in adequate amounts (Committee on Nutrition, American Academy of Pediatrics, 1970).

II. BIOAVAILABILITY OF IRON IN INFANT FORMULAS

A. Milk-Based Infant Formulas

Infant formulas based on milk (in the form of whole milk or skim milk) historically have been fortified with ferrous sulfate. The experimental justification for this practice includes the work of Niccum *et al.* (1953), who added

ISBN 0-12-177060-5

ferrous sulfate or ferric ammonium citrate to modified whole milk and found that infants fed the milk fortified with ferrous sulfate had higher hemoglobin levels than did infants fed the milk fortified with the ferric salt.

Numerous clinical comparisons of infants fed a milk-based formula with or without supplemental iron added as ferrous sulfate also showed that the added iron gave higher hemoglobin levels (Gross, 1970; Marsh *et al.*, 1959; Andelman and Sered, 1966; Gorton and Cross, 1964).

We (Theuer *et al.*, 1973) evaluated the bioavailability of several iron salts incorporated into sterilized liquid milk-based infant formula. Hemoglobin generation in anemic rats was the criterion of bioavailability. When incorporated into milk-based formulas, seven of eight salts, including ferrous sulfate, gave relative iron availabilities greater than standard ferrous sulfate (Table I). Standard ferrous sulfate was reagent grade ferrous sulfate mixed in a dry state to a standard casein–glucose diet.

The most significant finding in our study was that processing liquid infant formula greatly improved the bioavailability of iron in two salts (ferric pyrophosphate and sodium iron pyrophosphate) with poor to mediocre native bioavailability (Table II). We were able to show that the process of sterilization was responsible for this improvement and not the step of dissolution or dispersion of the iron salt in the liquid formula (Table II). Wood *et al.* (1978) have confirmed our observations on the beneficial effect of processing on the bioavailability of the iron of ferric pyrophosphate and sodium iron pyrophosphate in infant formula.

Table I
AVAILABILITY OF IRON IN STERILIZED LIQUID INFANT FORMULA

	Relative iron availability[a]	
Iron salt	Milk-based formula	Soy isolate formula
Ferrous sulfate	129	101
Ferrous citrate	148	89
Ferric citrate	122	87
Ferric gluconate	139	79
Ferrous lactate	118	100
Ferric glycerol phosphate	135	92
Ferric pyrophosphate	125	92
Sodium iron pyrophosphate	60	71

[a]Percentage of standard ferrous sulfate.

Table II

EFFECT OF PROCESSING ON AVAILABILITY OF IRON SALTS IN MILK-BASED INFANT FORMULA

Iron salt	Added to liquid formula	Formula sterilized with added iron	Relative iron availability[a]
Ferrous sulfate	No	No	126
	Yes	No	114
	Yes	Yes	129
Ferric pyrophosphate	No	No	78
	Yes	No	71
	Yes	Yes	125
Sodium iron pyrophosphate	No	No	42
	Yes	No	39
	Yes	Yes	60

[a]Percentage of standard ferrous sulfate.

B. Soy Isolate Infant Formulas

As noted elsewhere in this book by Coccodrilli and Shah (Chapter 11), adding iron to milk can give a gray cast to the product. This also is true for soy isolate infant formulas. The original soy isolate formula, introduced in 1965, contained sodium iron pyrophosphate. In extensive premarketing animal studies, no evidence of iron undernutrition was observed (R. Theuer and H. Sarett, unpublished studies).

However, Fritz *et al.* (1970) reported extremely poor bioavailability of this iron salt, so we (Theuer *et al.*, 1971) evaluated the bioavailability of this and seven other iron salts incorporated into liquid soy isolate infant formula (Table I). Formulas made with the eight salts had relative bioavailability ranging between 71 and 101% of standard ferrous sulfate added to the casein–glucose diet.

Soy isolate itself contains appreciable iron with good bioavailability (86% of standard ferrous sulfate), so we calculated the relative iron availability from the fortified formula as well as the availability from the iron salts used for fortification. Similar to our findings with milk-based formulas, processing significantly improved the relative bioavailability of ferric pyrophosphate and sodium iron pyrophosphate, two salts with poor to mediocre native bioavailability (Table III).

Interestingly, the relative bioavailability of the iron in soy isolate formulas was consistently less than that of the iron in milk-based formulas made with the same iron salt (Table I).

Table III
EFFECT OF PROCESSING ON AVAILABILITY OF IRON SALTS IN SOY ISOLATE INFANT FORMULA

Iron salt	Added to liquid formula	Formula sterilized with added iron	Relative availability of added iron[a]
Ferrous sulfate	No	No	90
	Yes	Yes	106
Ferric pyrophosphate	No	No	39
	Yes	Yes	93
Sodium iron pyrophosphate	No	No	15
	Yes	Yes	66

[a]Percentage of standard ferrous sulfate.

III. CONCLUSION

Infant formulas fortified with iron are relied on as an effective means of supplying iron to young infants. Laboratory evaluations confirm clinical observations that the iron in these products is highly bioavailable.

REFERENCES

Andelman, M. B., and Sered, B. R. (1966). Utilization of dietary iron by term infants. A study of 1,048 infants from a low socioeconomic population. *Am. J. Dis. Child.* **111,** 45–55.

Committee on Iron Deficiency, Council on Foods and Nutrition, American Medical Association. (1968). Iron deficiency in the United States. *JAMA, J. Am. Med. Assoc.* **203,** 61.

Committee on Nutrition. (Dec. 1970). ''Iron Fortified Formulas,'' Newsl. Suppl. American Academy of Pediatrics, Evanston, Illinois.

Committee on Nutrition, American Academy of Pediatrics. (1969). Iron balance and requirements in infancy. *Pediatrics* **43,** 134–142.

Fritz, J. C., Pla, G. W., Roberts, T., Boehne, J. W., and Hove, E. L. (1970). Biological availability in animals of iron from common dietary sources. *J. Agric. Food Chem.* **18,** 647–651.

Gorton, M. K., and Cross, E. R. (1964). Iron metabolism in premature infants. II. Prevention of iron deficiency. *J. Pediatr.* **64,** 509.

Gross, S. (1970). Relationship between the protein and iron content of cow's milk and hematological values in infancy. *Rep. Ross Conf. Pediatr. Res.* **62,** 89–101.

Marsh, A., Long, H., and Stierwalt, E. (1959). Comparative hematological response to iron fortification of a milk formula for infants. *Pediatrics* **24,** 404–412.

Niccum, W. I., Jackson, R. L., and Stearns, G. (1953). Use of ferric and ferrous iron in the prevention of hypochromic anemia in infants. *Am. J. Dis. Child.* **86,** 553–567.

Owens, G. M., Lubin, A. H., and Garry, P. J. (1971). Preschool children in the United States: Who has iron deficiency? *J. Pediatr.* **79,** 563–568.

Siimes, M. A., Salmenpera, L., and Perheentupa, J. (1984). Exclusive breast-feeding for 9 months: Risk of iron deficiency. *J. Pediatr.* **104,** 196–199.

Theuer, R. C., Kemmerer, K. S., Martin, W. H., Zoumas, B. L., and Sarett, H. P. (1971). Effect of processing on availability of iron salts in liquid infant formula products. Experimental soy isolate formulas. *J. Agric. Food Chem.* **19,** 555–558.

Theuer, R. C., Martin, W. H., Wallander, J. F., and Sarett, H. P. (1973). Effect of processing on availability of iron salts in liquid infant formula products. Experimental milk-based formulas. *J. Agric. Food Chem.* **21,** 482–485.

Wood, R. H., Stake, P. E., Eiseman, J. H., Shippee, R. L., Wolski, K. E., and Koehn, U. (1978). Effects of heat and pressure on the relative biological availability of selected dietary supplemental inorganic iron salts as determined by chick hemoglobin repletion assay. *J. Nutr.* **108,** 1477–1484.

10

Supplementation of Infant Products

GEORGE A. PURVIS
Gerber Research Center
Gerber Products Company
Fremont, Michigan

I. INTRODUCTION

The importance of supplemental iron for the infant is well recognized and has been the subject of extensive evaluation, development, and recommendation. Supplemental iron is imperative to minimize the risk of iron deficiency in the infant. The infant and young are the most important population group because of the expansion of blood volume and extremely rapid growth common during infancy and early childhood. When neonatal iron stores have been depleted—after about 2 months in small, preterm infants and 4–6 months in term infants—additional iron is essential. Breast milk contains small but significant amounts of iron, which is well utilized (American Academy of Pediatrics/Committee on Nutrition, 1976) and therefore sufficient for the breast-feeding period. Rapid growth, replacement of neonatal stores, and expansion of blood volume are critical contributors to iron requirements from the period 4–6 months to approximately 24 months.

Population groups have been identified that are particularly susceptible to iron deficiency during infancy and early childhood. A portion of the risk can be related to economic and social circumstances; however, a much greater problem exists because of failure to recommend iron-supplemented foods. Practices vary

ISBN 0-12-177060-5

significantly according to population group. Variations are also substantial on an international scale, since products, practices, and recommendations are variable. The heterogeneity of practices prevents a generalized discussion; however, iron-supplemented foods are available in most societies. Most references relate to practices in North America and Western Europe, since the information from these areas is well documented. Iron-supplemented foods are also universally available without restriction.

The established recommendations in North America have been presented by the American Academy of Pediatrics and the Canadian Paediatric Society (Canadian Paediatric Society Nutrition Committee, 1979). Practices in Western Europe have been defined by the European Society for Paediatric Gastroenterology. Foods and practices vary, but in all cases iron-supplemented foods are available. Regulation and recommendations recognize and accommodate those products.

II. RECOMMENDED SUPPLEMENTATION METHODS

The importance of feeding practices during the age period 6–24 months has been the subject of numerous evaluations, and strong recommendations have resulted. Practices in the United States and Canada indicate that more favorable iron nutrition has resulted from consumption of greater quantities of iron-supplemented foods (Sarett *et al.*, 1983; Johnson *et al.*, 1981; Yeung *et al.*, 1981). Iron-supplemented infant formula provides a reliable source of iron during the first year, with an appropriate complement by infant cereal during the period 3–24 months of age. The combination of these supplemented foods can assure favorable iron nutrition and at the same time allow for the individual variability in infants. The contribution to total nutrition of infants by the supplemented foods accounts for disproportionately less calorie content than iron.

A. Ferrous Sulfate in Infant Formulas

Iron supplementation of infant formulas in the United States and Canada is in the form of ferrous sulfate. The availability of iron in infant formulas from this source has been established (Theuer *et al.*, 1971, 1973) for both milk-based and soy-based formulas. Labeling clearly identifies "infant formulas with iron." There are formula counterparts prepared with no added iron. The amount of iron in iron-supplemented formulas varies within a reasonable range with a Codex minimum established at 1 mg/100 kcal (Codex Alimentarius Commission, 1976).

B. Effect of Cow's Milk versus Breast-Feeding on Iron Requirement

The introduction of whole cow's milk into the infant diet distinctly influences the iron nutrition for the infant. The iron content of cow's milk is extremely low, and, as a result, an obvious dilution occurs when cow's milk replaces other more iron-rich sources. In addition to its low iron concentration, cow's milk interferes with iron absorption both by affecting bioavailability of other dietary sources and by causing occult blood loss (Eastham and Walker, 1977). Iron-supplemented food sources with demonstrated efficacy are available and can assure beneficial iron nutrition for the infant and young child. Recommendations are not in total agreement on this subject. Those recommendations that are oriented to breast-fed populations recommend breast-feeding with ''appropriate supplementation'' during the 4- to 12-month age period. More specific iron recommendations in North America suggest a combination of iron-rich foods and heat-treated milk sources (precooked iron-supplemented infant cereals and iron-supplemented infant formula) (American Academy of Pediatrics/Committee on Nutrition, 1976). The practice in Western Europe suggests supplemented infant formula as the most reasonable iron source during the period after 4–6 months of age because of frequency of consumption (ESPGAN Committee on Nutrition, 1981). Heat-treated cow's milk or, preferably, iron-supplemented infant formula (which is also heat treated) is clearly the preferred milk source. Infant cereal has demonstrated effectiveness as a food for early introduction as a supplement to breast-feeding or to infant formula. Several characteristics distinguish infant cereal as the most favorable food for infants:precooked, ready to feed, partially hydrolyzed, and iron supplemented.

III. FORMS OF IRON USED IN SUPPLEMENTATION

The iron compound preferred for supplementation of infant cereal has been designated as small-particle electrolytic iron (less than 375 mesh, particles less than 45 μm) (Fritz *et al.*, 1975). The favorable absorption of this compound (also ferrous sulfate in formula) has been established with infant-feeding experiments (Rios *et al.*, 1975), in addition to numerous animal studies. Electrolytic iron is used in the vast majority of precooked infant cereal products; it does not discolor cereal, does not induce flavor deterioration, and is the most favorably absorbed form. Ready-to-feed cereal with fruit in jars is supplemented with ferrous sulfate. Experiments are being continued with alternate forms; however, present formulas are widely acknowledged to be excellent.

There are other acceptable supplementary forms of iron. European preference includes carbonyl iron and iron ethylenediaminetetraacetic acid (EDTA). Carbonyl iron can be prepared in extremely small particle size and represents a processed form of elemental iron, as does electrolytic iron.

Several iron forms have promise for more extensive future use. Attributes include good availability, stability in food products, and lack of interaction with other food components. Ferrous salts of EDTA have demonstrated excellent absorptive characteristics (Viteri *et al.*, 1978). EDTA salts are also compatible with many kinds of processed foods, since they are relatively stable in the temperature, pH, and moisture ranges normally encountered. EDTA salts are commonly used in Europe. Since some forms, particularly ferric forms, are not well absorbed and since EDTA is recognized as a strong sequestrant, acceptance has not been extensive in North America. Ferric fructose compounds have also sparked great interest as they appear to be more efficiently absorbed than ferrous sulfate—up to three times greater retention in guinea pigs (Bates *et al.*, 1972). The mechanism for increased absorbance is unknown but may involve reduction of the ferric ion or the metabolism of fructose, enhancing iron uptake by intestinal cells. Studies of iron availability should be considered carefully in regard to procedural differences that can greatly alter the results of such experiments.

Sodium iron pyrophosphate was utilized for supplementation of infant cereal for several years as the result of an observation in 1958 that absorption was favorable (Schultz and Smith, 1958). The absorption was not confirmed in later studies (Rios *et al.*, 1975) and has not been corroborated with animal tests. Sodium iron pyrophosphate should not be used for supplementation.

IV. RECOMMENDED AMOUNTS OF IRON IN SUPPLEMENTED FOODS

The amounts of iron in products for infants have been established by theoretical design followed by usage in most instances. A universal set of minima for iron supplementation has been recommended for infant formula with iron at 1 milligram per 100 available kilocalories (Codex Alimentarius Commission, 1976). The ranges for usage that exist in the United States and Canada are the following: infant formula, 10–15 mg/liter iron as ferrous sulfate; infant cereal, 0.47 mg elemental iron per gram of cereal; and 4.5–6.75 mg iron as ferrous sulfate in ready-to-feed cereal with fruit. These quantities translate to reasonable dietary intakes; 1 liter of formula is considered a reasonable maximum intake. Reasonable infant cereal consumption ranges from 7 to 14 g/serving, which is equivalent to 3–6 mg iron/serving.

There are several cereal-based products frequently offered to infants such as farina and other cream cereals. Because of the varying amounts of fortification and the variety of iron forms used, these are not considered a reliable iron source.

Most are wheat based, and due to the lack of favorable tolerance by infants, wheat introduction should be delayed past the first year. The prolamine content may relate to characteristics of intolerance. Cereal–fruit–formula mixtures are also marketed with a limited iron content—approximately 3 mg/28-g serving—and are not considered a significant contributor to iron nutrition.

V. SUMMARY AND CONCLUSIONS

The combination of iron-supplemented foods has improved the iron nutrition of infants in the United States and Canada during the past decade, serving as an indication that a combination of supplementation practice, sound recommendation, and effective monitoring lead to favorable intakes. All factors are essential to avoid iron deficiency. The total diet must be considered, since absorption of iron from various foods is inconsistent. The absorption of iron from food has been reported to range from 1 to more than 20%. The form of iron has well-defined importance, as has the iron nutrition state of the infant. Heme iron from meat and selected other animal sources not only is well absorbed but also enhances the iron absorption from the remainder of the food consumed. Interference by milk is attributable to the form of iron present as well as the iron content, since the iron present is not in the form of heme iron. Ascorbic acid enhances nonheme iron absorption from foods. On the other hand, certain naturally occurring sequestrants in food tend to interfere with iron absorption; tea and coffee have been noted (Mahoney and Hendricks, 1984).

The importance of iron in the infant diet clearly represents a nutrition problem that can be avoided through appropriate dietary intake, monitoring, and availability of iron-supplemented foods. Alternative provision should be made for preterm infants and for infants breast-fed for an extended period, since the reduced caloric intake and rapid growth require greater iron concentration. Iron-supplemented infant formula and iron-supplemented infant cereal have proved effective in the provision of adequate quantities of readily available iron. Ferrous sulfate can be satisfactorily used in infant formula and other foods that do not contain reactive forms or in which conditions are favorable (pH, moisture, heat treatment). Finely divided elemental iron is effective in reactive food systems and can be used for foods which are not compatible with ferrous sulfate.

REFERENCES

American Academy of Pediatrics/Committee on Nutrition. (1976). Iron supplementation for infants *Pediatrics* **58,** 765.

Bates, G. W., Boyer, J., Hegenauer, C., and Saltman, P. (1972). Facilitation of iron absorption by ferrous fructose. *Am. J. Clin. Nutr.* **25,** 983–986.

Canadian Paediatric Society Nutrition Committee (1979). Infant feeding. *Can. J. Public Health* **70,** 376.

Codex Alimentarius Commission (1976). "Recommended International Standards for Foods for Infants and Children," CAC/RS 72/74. Food and Agriculture Organization, Rome.

Eastham, E. J., and Walker, W. A. (1977). Effect of cow's milk on the gastrointestinal tract: A persistent dilemma for the pediatrician. *Pediatrics* **60,** 477.

ESPGAN Committee on Nutrition (1981). Guidelines on infant nutrition. *Acta Paediatr. Scand., Suppl.* **287.**

Fritz, J. C., Pla, G. W., and Rolinson, C. L. (1975). Iron for enrichment. *Baker's Dig.* **49,** 46.

Johnson, G. H., Purvis, G. A., and Wallace, R. D. (1981). What do infants really eat. *Nutr. Today* **16,** 4.

Mahoney, A. W., and Hendricks, D. G. (1984). Potential of the rat as a model for predicting iron bioavailability for humans. *Nutr. Res.* **4,** 913.

Rios, E., Hunter, R. E., Cook, J. D., Smith, N. J., and Finch, C. A. (1975). The absorption of iron as supplements in infant cereal and infant formulas. *Pediatrics* **55,** 686.

Sarett, H. P., Bain, K. R., and O'Leary, J. C. (1983). Decisions on breast-feeding or formula-feeding and trends in infant-feeding practices. *Am. J. Dis. Child.* **137,** 719.

Schultz, J., and Smith, N. J. (1958). Quantitative study of the absorption of iron salts in infants and children. *Am. J. Dis. Child.* **95,** 120–125.

Theuer, R. C., Kemmerer, K. S., Martin, W. H., and Sarett, H. P. (1971). Effect of processing on availability of iron salts in liquid infant formula products. Experimental soy isolate formulas. *J. Agric. Food Chem.* **19,** 555.

Theuer, R. C., Martin, W. H., Wallander, J. F., and Sarett, H. P. (1973). Effect of processing on availability of iron salts in liquid infant formula products. Experimental milk-based formulas. *J. Agric. Food Chem.* **21,** 482.

Viteri, F. E. *et al.* (1978). Sodium iron Na Fe EDTA as an iron fortification compound in Central America—absorption studies. *Am. J. Clin. Nutr.* **31,** 961–971.

Yeung, D. L., Pennell, M. D., Leung, M., Hall, J., and Anderson, G. H. (1981). Iron intake of infants; the importance of infant cereals. *Can. Med. Assoc. J.* **125,** 999.

11

Beverages

G. COCCODRILLI, JR.
N. SHAH*

Nutrition and Health Sciences
General Foods Corporation
Technical Center
Tarrytown, New York

I. INTRODUCTION

The variety of beverages consumed throughout the world could be used as an effective and convenient form of iron fortification in the human diet. Beverages are consumed by identifiable population segments. For example, coffee and tea are normally consumed by adults; milk, fruit and vegetable juices, fruit- and vegetable-flavored drinks, and soft drinks are commonly consumed by all age groups. Iron is well absorbed from most beverages, and consumption of iron-fortified beverages between meals can increase iron absorption three- to fourfold over absorption from a vegetal meal, as shown by Layrisse *et al.* (1976a,b). Some of these beverages can also be fortified with other nutrients, including ascorbic acid, an enhancer of iron absorption, as shown by Sayers *et al.* (1973, 1974a,b), Cook and Monsen (1977), Derman *et al.* (1980), Rossander *et al.*

*Present address: Nutrition and Health Sciences, General Foods Corporation, Cranbury Technical Center, Cranbury, New Jersey.

ISBN 0-12-177060-5

(1979), Bjorn-Rasmussen and Hallberg (1974), and Hallberg and Rossander (1982) among many others.

The primary consideration in developing a beverage system as a vehicle for iron fortification is identification of a biologically available iron source that is compatible with the product from manufacture to consumption and does not alter organoleptic properties of the product. A series of criteria to be considered in developing an iron-fortified beverage are summarized in Table I.

Iron fortification of milk and milk-based beverages, of coffee, of fruit- and vegetable-flavored drinks, and of soft drinks will be discussed.

Table I

SELECTED CRITERIA FOR DEVELOPMENT OF IRON-FORTIFIED BEVERAGES

Identification of food-approved iron compounds
Bioavailability
Cost
Color
Solubility
Particle size
Commercial availability
Manufacturing practices and product form
Heat processing
Pasteurization, sterilization, spray-drying, evaporation–condensation
Freeze-dried powders
Instant powdered beverage mixes
Method of addition and nutrient dispersibility
Dry blending—premixes, spray coatings
Product stability
Packaging
Foil pouch, can, paperboard, plastic, glass, paper envelope
Environmental conditions
Climate, temperature, humidity
Expected shelf life
Interactions with flavor and color systems
Interactions with other nutrients
Target consumer group
Serving size
Frequency of anticipated consumption
Fortification level (mg Fe/serving)
Labeling regulations (marketing claims of ''fortified with iron'')

II. MILK AND MILK-BASED PRODUCTS

Universality of consumption of milk and milk products makes them potentially good vehicles for delivering iron in the diet. However, whole milk has a unique and delicate series of flavor notes, and even minor alterations by addition of iron compounds may result in an organoleptically inferior product. In fact, notable progress has been made in dairy processing to eliminate metal contaminants; therefore, it is a bit ironic that we are today attempting to fortify milk with metals.

Addition of ferrous and ferric salts to whole milk enhances some or all of the following reactions:

1. Lipolytic rancidity caused by action of lipases present in milk
2. Oxidative rancidity caused by oxidation of free fatty acids, giving rise to "oxidized" flavor
3. Partial or complete loss of nutrients such as vitamin A, carotene, and ascorbic acid

Iron fortification of whole milk, of evaporated, skim, and powdered milk, and of milk-based beverages will be discussed. In judging some of the contradictory data presented here, it should be noted that there are tremendous variations in milks due to season, breed of cows, feeding practices for cows, geographical origin, analytical variance of nutrient analyses, and regulations regarding percentage of fat and nonfat solids.

A. Pasteurized Whole Milk

The following iron compounds have been assessed in the fortification of pasteurized whole milk: ferric ammonium citrate, ferric ammonium sulfate, ferric choline citrate, ferric pyrophosphate, ferric phosphate, ferric glycerophosphate, ferric citrate, ferric lactate, ferrous sulfate, ferrous gluconate, ferrous lactate, ferrous ammonium sulfate, and ferrous fumarate.

Edmondson *et al.* (1971) showed that use of ferric compounds in whole milk resulted in lipolytic rancidity when milk was pasteurized below 79°C. This off-flavor was reduced or eliminated by pasteurizing at 81°C. Data showed that milk lipase is made more heat resistant by ferric compounds. Ferrous compounds caused oxidized flavor when added to milk before pasteurization. This off-flavor was markedly reduced by de-aerating milk before adding iron. Ferrous compounds did not cause rancidity. Ferric ammonium citrate addition followed by pasteurization at 81°C was judged the most desirable procedure for fortifying milk.

Wang and King (1973a) also found ferric ammonium citrate to be the compound of choice in fortifying milk with iron. Ferric ammonium citrate up to 30 ppm (ppm = mg/kg) iron in milk stored for up to 7 days was acceptable in organoleptic testing by trained panelists and consumers; and the contents of vitamin E, vitamin A, and carotene were not reduced.

DeMott (1971) tested ferric pyrophosphate, ferric phosphate, ferric ammonium citrate, ferrous sulfate, ferrous gluconate, and ferrous lactate at 5.2–12.6 ppm iron in milk. All the compounds caused off-flavor in milk and skim milk, but the effects of ferric pyrophosphate and ferric phosphate were only minor. Soluble ferric pyrophosphate was identified as the ingredient in a multivitamin mineral preparation causing the oxidized flavor in homogenized pasteurized milk by Scanlan and Shipe (1962).

Although ascorbic acid is known to enhance iron absorption, as mentioned earlier, its combination with iron causes problems due to the susceptibility of ascorbic acid to oxidation by iron salts. Hegenauer *et al.* (1979a) showed an increase only in the rate of oxidation of ascorbate to dehydroascorbate due to iron, and not in the equilibrium between ascorbate, dehydroascorbate, and diketogulonate. The conversion of ascorbate to diketogulonate occurred rapidly even in the unsupplemented milk. Thus, iron addition may not affect the ascorbic acid content of milk.

Kiran *et al.* (1977) have used ferric lactose with ascorbic acid in milk at 125 ppm iron. No appreciable detrimental effect could be observed on color, pH, or organoleptic flavor of milk. When ferric lactose was used alone, a slight color change was observed in the samples.

Ranhotra *et al.* (1981) tested bioavailability of citrate phosphate iron complex at 38 ppm iron in pasteurized homogenized milk. They found that milk does not adversely affect bioavailability of added iron. In an iron depletion–repletion study in rats, iron from citrate phosphate iron complex was as available as iron from ferrous sulfate.

Wang and King (1973b) studied iron assimilation from milk fortified with ferric ammonium citrate in baby pigs. About 30% of the dietary iron was absorbed, and 90% of that was incorporated in red blood cells.

To minimize reactions leading to oxidative deterioration in iron-fortified milk, Hegenauer *et al.* (1979b,c,d) have recommended the addition of Fe(III) chelates of nitrilotriacetate and lactobionate to homogenized milk, just before pasteurization. They postulate that Fe(III) chelates bind rapidly to casein and insignificantly to milk fat, thus protecting milk from oxidation. However, these compounds are not currently practical alternatives, in that they are not approved for food use, and a commercial supplier is not available. In a human clinical trial, ferric lactobionate added to milk was shown to have poorer bioavailability than ferrous ion in the presence of copper by Rivera *et al.* (1982).

In addition to the compounds previously described, Baldwin *et al.* (1982) has evaluated an iron complex of citric acid–potassium hydroxide–phosphoric acid–reduced iron in pasteurized whole milk. Although a slight oxidized flavor was detectable in pasteurized whole-milk samples, the iron complex (~15% iron) was rated by an experienced taste panel as having significant potential for use in iron supplementation of milk.

B. Evaporated, Skim, and Powdered Milk

Evaporated milk products are vigorously heat treated, and the resultant products are quite different from whole milk. Owen and McIntire (1975) tested 11 iron compounds for compatibility in an evaporated milk product. When added at the levels of 11, 22, and 42 ppm iron, only one compound, ferric orthophosphate, showed usefulness. Poor solubility of ferric orthophosphate makes even distribution of iron in evaporated milk a technological problem. They also reported that ferrous sulfate added by dry blending to nonfat dry milk at 10 ppm iron was acceptable by taste panel evaluations following a 12-month storage period at ambient temperatures.

Kurtz *et al.* (1973) evaluated effects of fortifying skim milk, skim milk concentrate, and nonfat dry milk powder with ferric ammonium citrate and ferric chloride at 20 ppm iron in constituted milk. No adverse flavor effects were observed. The only adverse effect of iron-fortified nonfat dry milk powder was an undesirable color change when it was added to cocoa, tea, or coffee. In rat studies, iron from these iron-fortified milks was as available as that from ferrous sulfate.

Schoppet *et al.* (1974) described a process of vacuum foam-drying of whole milk fortified with ferric ammonium citrate at 10.6 ppm iron. There were no differences in organoleptic properties of the fortified and the nonfortified samples.

Layrisse *et al.* (1973) studied iron absorption from skim milk enriched with iron glycerophosphate in preschool children in Venezuela. Iron absorption from iron glycerophosphate and ferrous sulfate were similar.

Davis *et al.* (1976) found that iron absorption from ferric ammonium citrate added to reconstituted whole milk powder was approximately 10% in humans, making this milk preparation a good source of iron.

C. Milk-Based Beverages

Douglas *et al.* (1981) tested chocolate milks fortified with all the iron compounds listed for fortification of whole milk, plus sodium ferric pyrophosphate, ferripolyphosphate, and ferripolyphosphate–whey protein complex for color and flavor. Sodium ferric pyrophosphate, ferripolyphosphate, and ferripolyphosphate–whey protein complex produced little or no off-color change in samples stored for up to 2 weeks. All the other compounds produced large color changes. Ferric compounds were judged better than ferrous compounds in keeping the original chocolate milk flavor, but the flavor scores for the samples fortified with the ferrous compounds improved after storage for 14 days at 4°C. In bioavailability testing in rats, ferripolyphosphate–whey protein complex was utilized as well as ferrous sulfate, but sodium ferric pyrophosphate was utilized only 35% as well as ferrous sulfate.

Effects of long-term storage of an iron-fortified canned milk-based product on relative biological value (RBV) of iron in rats were studied by Clemens (1981). The samples were fortified with ferric orthophosphate, electrolytic iron, or carbonyl iron at 25% of USRDA for iron. The RBVs for the two elemental iron sources were comparable to the RBV of ferrous sulfate and were significantly greater than the RBV for ferric orthophosphate at the 6-month and the 12-month storage periods. A markedly greater percentage of iron from the two elemental iron sources than from ferric orthophosphate was solubilized and found in the ferrous form.

III. COFFEE

Coffee is consumed primarily by adult populations throughout much of the world, and therefore may serve as an appropriate vehicle for iron fortification. Discussion will be limited to the two major coffee products: instant soluble, and roast and ground.

A. Instant Soluble Coffee

Klug *et al.* (1977) identified a process for iron fortification of a soluble coffee product that overcame the two primary problems associated with iron addition: off-color and formation of sediment in the cup. Addition of soluble iron to coffee imparts a greenish tint, which is most noticeable on addition of cream or whitener. Sedimentation is the result of reactions between elemental iron and polyhydroxyphenols and polyhydroxyphenol–polysaccharide materials present in the coffee.

By tempering the coffee extract at approximately 21°C, followed by centrifugation, most of the iron-reactive compounds are removed. The resulting extract is then fortified with soluble ferric pyrophosphate, and subjected to low-temperature spray-drying and agglomeration. All soluble ferrous and ferric salts can be used in a wide range of fortification levels (0.01 to 1.0 parts elemental iron per 100 parts soluble coffee solids). Bioavailability testing of this product in our laboratory showed no significant interference with iron absorption using a rat bioassay and comparing the response of ferric pyrophosphate to ferrous sulfate.

B. Roast and Ground Coffee

Iron fortification is more difficult in roast and ground coffee than in soluble coffee because of variations in brew strength and in methods of preparation (e.g., perk versus drip).

Following the roasting and grinding of coffee beans, iron compounds may be added to a dry-blending process before vacuum packaging. Several commercially available soluble iron salts have potential in terms of color and solubility compatibility. Johnson and Evans (1977) have reported on the use of ferrous fumarate, a water-soluble reddish brown compound with a good bioavailability rating, in roast and ground coffee with iron addition at 5 ppm iron in brewed coffee. Twenty laboratory people who tasted the fortified and the nonfortified coffee could not find any difference in the organoleptic properties of the two. Bioassay of this product for iron in rats showed an average absorption value of 31.8 ± 11.9% for black coffee, and the addition of milk, sugar, or nondairy creamer did not significantly alter the iron absorption.

At the level of 5 ppm iron, one cup of coffee (200 ml) provides 1 mg of iron. If iron is absorbed from coffee at the 5–15% absorption level commonly cited for nonheme sources, coffee, depending on the amount consumed, could supply a substantial percentage of the daily requirement of iron. It is important to note that consumption patterns of a product should be carefully assessed before embarking on any fortification effort to avoid the potential hazard of excessive intakes.

IV. FRUIT- AND VEGETABLE-FLAVORED BEVERAGES

Fruit and vegetable juices and fruit drinks are widely consumed by population groups worldwide. Most of these beverages are good sources of ascorbic acid, which is either inherent to the product or added as a supplemental nutrient. Reports by Sayers *et al.* (1973, 1974a,b), Cook and Monsen (1977), Derman *et al.* (1980), Rossander *et al.* (1979), Bjorn-Rasmussen and Hallberg (1974), and Hallberg and Rossander (1982) among many others have shown that ascorbic acid is probably one of the most effective ways of increasing the bioavailability of iron in a meal. Thus, fortification of these beverages containing a known enhancer of iron absorption (i.e., ascorbic acid) with iron seems to be a prudent and rational approach in attempting to reduce the worldwide problem of iron deficiency anemia. Powdered beverage mixes can also play an important role, especially in developing countries where refrigerated distribution systems might be lacking.

Beverages are sold in a variety of forms, as ready-to-drink, as frozen concentrates, or as powdered mixes. Regardless of the form of the beverage, the addition of iron can cause the following problems:

1. Some of the ascorbic acid is destroyed.
2. The presence of vitamins such as thiamine, vitamin A, folic acid, and ascorbic acid causes flavor and taste deterioration.
3. Iron addition alone at levels exceeding 15% of USRDA (2.7 mg iron/serving) can be detected as a metallic taste in some beverage systems.
4. Iron addition sometimes decolorizes the product.

Morse and Hammes (1976) described a procedure of stabilizing ascorbic acid and iron in bottled and canned beverages by adding cysteine in specific ratios. Storage studies showed no adverse effect of cysteine on taste. There was some destruction of ascorbic acid depending on the storage conditions and the beverage types.

In our laboratory we have experimented with the fortification profile for powdered breakfast drink mixes to include iron, vitamin A, and ascorbic acid. Anhydrous ferrous sulfate was selected for fortification from an extensive screening of food grade iron compounds that would be stable in the dry powder mix, and soluble in a low-pH, high-acid beverage on reconstitution. Ferrous sulfate was found suitable for attaining fortification levels up to 15% of USRDA (2.7 mg iron/serving). All samples were packed in foil pouches and subjected to accelerated storage conditions to emulate shelf conditions of high temperature (90°F) and high relative humidity (70%). Intensive taste panel evaluations identified a slightly different flavor profile for the iron-fortified product, but the product was judged acceptable in all the evaluations. Since vitamin A and ascorbic acid were common in both the control and the iron-fortified samples, the slight taste change in the iron-fortified product was due to the addition of iron.

An iron bioavailability test using a slope ratio bioassay measuring hemoglobin regeneration in iron-depleted rats was performed on the iron-fortified product. Results indicated that even after accelerated storage for 4 months, there was no adverse effect on the bioavailability of iron. In our recent work, ferrous gluconate was found to be organoleptically superior to ferrous sulfate in similar beverage systems.

V. SOFT DRINKS

A paucity of information exists in the technical literature on attempts to fortify soft drinks. This is attributable in part to the decision by various trade associations and manufacturers to treat the product category usually as a refreshment beverage, thereby not assigning any specific dietary role of nutrient delivery to this product category. Some work has been reported by Layrisse *et al.* (1976a,b) in which carbonated soft drinks and orange juice were formulated with iron-fortified sucrose and tested in human clinical trials in Venezuela. Advantages of using sugar as a vehicle of iron fortification include convenience of solubility of sugar in beverages and absence of inhibitors of iron absorption in it. Iron from sugar fortified with ferrous sulfate or with Fe(III) EDTA complex was absorbed at similar levels. As mentioned earlier, consumption of beverages containing iron-fortified sugar between meals increased iron absorption three- to fourfold over that from a vegetal meal.

VI. SUMMARY

Iron deficiency anemia is a significant worldwide public health problem. With proper constraint and concern for the issue of iron overload, it is possible to use beverages as an effective and convenient diet item for delivery of bioavailable iron.

REFERENCES

Baldwin, R. E., Shelley, D. S., and Marshall, R. T. (1982). Flavor of milk one week after addition of iron complex. *J. Dairy Sci.* **65,** 1390–1393.

Bjorn-Rasmussen, E., and Hallberg, L. (1974). Iron absorption from maize. Effect of ascorbic acid on iron absorption from maize supplemented with ferrous sulphate. *Nutr. Metab.* **16,** 94–100.

Clemens, R. A. (1981). Effects of storage on the bioavailability and chemistry of iron powders in a heat-processed liquid milk-based product. *J. Food Sci.* **47,** 228–230.

Cook, J. D., and Monsen, E. R. (1977). Vitamin C, the common cold and iron absorption. *Am. J. Clin. Nutr.* **30,** 235–241.

Davis, A. E., Bolin, T. D., and Callender, S. T. (1976). The bio-availability of an iron fortified milk powder. *Nutr., Proc. Int. Congr., 10th, 1975* pp. 219–220.

DeMott, B. J. (1971). Effects on flavor of fortifying milk with iron and absorption of the iron from intestinal tract of rats. *J. Dairy Sci.* **54,** 1609–1614.

Derman, D. P., Bothwell, T. H., MacPhail, A. P., Torrance, J. D., Bezwoda, W. R., Charlton, R. W., and Mayet, F. (1980). Importance of ascorbic acid in the absorption of iron from infant food. *Scand. J. Haematol.* **25,** 193–201.

Douglas, F. W., Jr., Rainey, N. H., Wong, N. P., Edmondson, L. F., and LaCroix, D. E. (1981). Color, flavor, and iron bioavailability in iron-fortified chocolate milk. *J. Dairy Sci.* **64,** 1785–1793.

Edmondson, L. F., Douglas, F. W., Jr., and Avants, J. K. (1971). Enrichment of pasteurized whole milk with iron. *J. Dairy Sci.* **54,** 1422–1426.

Hallberg, L., and Rossander, L. (1982). Effect of different drinks on the absorption of non-heme iron from composite meals. *Hum. Nutr.: Appl. Nutr.* **36A,** 116–123.

Hegenauer, J., Saltman, P., and Ludwig, D. (1979a). Degradation of ascorbic acid (vitamin C) in iron-supplemented cows' milk. *J. Dairy Sci.* **62,** 1037–1040.

Hegenauer, J., Saltman, P., Ludwig, D., Ripley, L., and Bajo, P. (1979b). Effects of supplemental iron and copper on lipid oxidation in milk. 1. Comparison of metal complexes in emulsified and homogenized milk. *J. Agric. Food Chem.* **27,** 860–867.

Hegenauer, J., Saltman, P., and Ludwig, D. (1979c). Effects of supplemental iron and copper on lipid oxidation in milk. 2. Comparison of metal complexes in heated and pasteurized milk. *J. Agric. Food Chem.* **27,** 868–871.

Hegenauer, J., Saltman, P., Ludwig, D., Ripley, L., and Ley, A. (1979d). Iron-supplemented cow milk. Identification and spectral properties of iron bound to casein micelles. *J. Agric. Food Chem.* **27,** 1294–1301.

Johnson, P. E., and Evans, G. W. (1977). Coffee as a low-calorie vehicle for iron-fortification. *Nutr. Rep. Int.* **16,** 89–92.

Kiran, R., Amma, M. K. P., and Sareen, K. N. (1977). Milk fortification with a system containing both iron and ascorbic acid. *Indian J. Nutr. Diet.* **14,** 260–266.

Klug, S. L., Patrizio, F. J., and Einstman, W. J. (1977). Iron-fortified soluble coffee and method for preparing same. U.S. Patent 4,006,263.

Kurtz, F. E., Tamsma, A., and Pallansch, M. J. (1973). Effect of fortification with iron on susceptibility of skim milk and nonfat dry milk to oxidation. *J. Dairy Sci.* **56,** 1139–1143.

Layrisse, M., Martinez-Torres, C., Ruphael-Divo, M., Jaffe, W., and Torres-Suarez, J. E. (1973). Iron absorption from skin milk enriched with iron glycerophosphate. *Arch. Latinoam. Nutr.* **23,** 145–150.

Layrisse, M., Martinez-Torres, C., Renzi, M., Velez, F., and González, M. (1976a). Sugar as a vehicle for iron fortification. *Am. J. Clin. Nutr.* **29,** 8–18.

Layrisse, M., Martinez-Torres, C., and Renzi, M. (1976b). Sugar as a vehicle for iron fortification: Further studies. *Am. J. Clin. Nutr.* **29,** 274–279.

Morse, L. D., and Hammes, P. A. (1976). Beverage containing stabilized vitamin C and iron and method of making same. U.S. Patent 3,958,017.

Owen, D. F., and McIntire, J. M. (1975). Technologies of the fortification of milk products. *Technol. Fortification Foods, Proc. Workshop 1975* pp. 44–65.

Ranhotra, G. S., Gelroth, J. A., Torrence, F. A., Bock, M. A., and Winterringer, G. L. (1981). Bioavailability of iron in iron-fortified fluid milk. *J. Food Sci.* **46,** 1342–1344.

Rivera, R., Ruiz, R., Hegenauer, J., Saltman, P., and Green, R. (1982). Bioavailability of iron- and copper-supplemented milk for Mexican school children. *Am. J. Clin. Nutr.* **36,** 1162–1169.

Rossander, L., Hallberg, L., and Björn-Rasmussen, E. (1979). Absorption of iron from breakfast meals. *Am. J. Clin. Nutr.* **32,** 2484–2489.

Sayers, M. H., Lynch, S. R., Jacobs, P., Charlton, R. W., Bothwell, T. H., Walker, R. B., and Mayet, F. (1973). The effects of ascorbic acid supplementation on the absorption of iron in maize, wheat and soya. *Br. J. Haematol.* **24,** 209–218.

Sayers, M. H., Lynch, S. R., Charlton, R. W., Bothwell, T. H., Walker, R. B., and Mayet, F. (1974a). The fortification of common salt with ascorbic acid and iron. *Br. J. Haematol.* **28,** 483–495.

Sayers, M. H., Lynch, S. R., Charlton, R. W., Bothwell, T. H., Walker, R. B., and Mayet, F. (1974b). Iron absorption from rice meals cooked with fortified salt containing ferrous sulphate and ascorbic acid. *Br. J. Nutr.* **31,** 367–375.

Scanlan, R. A., and Shipe, W. F. (1962). Factors affecting the susceptibility of multivitamin mineral milk to oxidation, *J. Dairy Sci.* **45,** 1449–1455.

Schoppet, E. F., Panzer, C. C., Talley, F. B., and Sinnamon, H. I. (1974). Continuous vacuum foam-drying of whole milk. VI. Iron enrichment. *J. Dairy Sci.* **57,** 1256–1257.

Wang, C. F., and King, R. L. (1973a). Chemical and sensory evaluation of iron-fortified milk. *J. Food Sci.* **38,** 938–940.

Wang, C. F., and King, R. L. (1973b). Assimilation of iron from iron-fortified milk by baby pigs. *J. Food Sci.* **38,** 941–944.

12
Salt

B. S. NARASINGA RAO
National Institute of Nutrition
Indian Council of Medical Research
Hyderabad, India

I. SALT AS A VEHICLE FOR IRON FORTIFICATION

Iron fortification has been considered as one of the practical approaches for the prevention and control of iron deficiency anemia in the population (International Nutritional Anemia Consultative Group, 1977). In India and some other developing countries like Indonesia and Thailand, common salt is considered to be a suitable vehicle for iron fortification satisfying all the criteria of an ideal vehicle. In India,

1. Salt is consumed by all segments of the population, rich as well as poor, perhaps somewhat more by the poor.
2. Salt consumption lies within a narrow range of 12–20 g/day, with an average intake of 15 g/day.
3. Of the estimated annual production of 7 million tons of salt, about 4.5 million tons are estimated to be available for human consumption, most of which is produced in about 120 manufacturing centers.

ISBN 0-12-177060-5

A well-established salt distribution system exists in India. Most of the salt consumed in India is crystal salt manufactured from brine (seawater) and is relatively crude, containing as much as 4% moisture. Salt that is normally distributed from salt works has the following composition:

NaCl	94–95%
Water	4%
Ca sulfate	0.4%
Magnesium salt	0.5–1.6%
Insolubles	0.1%

The high moisture content and the presence of magnesium salts pose several technical problems in fortifying crude salt with iron.

II. IRON SOURCES FOR SALT FORTIFICATION

In identifying the iron source for fortifying salt, the following criteria were considered:

1. It must be stable when mixed with salt and not develop any color.
2. The bioavailability of iron in the fortified salt must be satisfactory, particularly when the iron-fortified salt is added to food, cooked, and consumed.
3. It must not impart color or taste to the food to which it is added.
4. It must be stable under the prevailing conditions of storage and transportation of salt.

A number of iron compounds were investigated singly or in combination with other chemical agents to achieve these objectives.

A. Soluble Compounds

Ferrous sulfate is the iron source of choice from the viewpoint of cost and bioavailability. When ferrous sulfate or other soluble iron compounds with good bioavailability were added to crude salt, the color of salt changed to brown or yellow either immediately or within a few days of mixing as a result of oxidation and hydrolysis of the iron compound. The iron compounds so tested are listed in Table I. None of the compounds tested was found to be satisfactory. Color development was slower with ferrous salts than with ferric salts.

B. Insoluble Compounds

Next, some of the insoluble iron sources were investigated. The reason for selecting these was twofold: (1) Some of them, like iron phosphates and reduced

Table I
SOLUBLE IRON SOURCES TESTED FOR SALT FORTIFICATION

Iron source	Color of fortified salt	Time of color development
Ferrous sulfate	Yellow	7 days
Ferrous ammonium sulfate	Yellow	7 days
Ferric ammonium sulfate	Yellow	$\leq$1 day
Ferric ammonium citrate	Brown yellow	$\leq$1 day
Ferric EDTA	Brown	$\leq$1 day

iron, had been used earlier for food fortification; and (2) insoluble iron compounds were expected to be stable when added to crude salt with a high moisture content. The following compounds were tested: ferric phosphate, ferric pyrophosphate, sodium iron pyrophosphate, monoferrous acid citrate, and reduced iron powder (Mallinckrodt). The first three salts just listed are white or buff-colored compounds that when mixed with salt were found to be highly stable for several months under a variety of storage conditions. After testing the stability of salt fortified with these iron phosphates, absorption of iron from them was tested in human volunteers using ^{59}Fe-labeled salts and whole-body counting. Iron absorption from these iron phosphates was unacceptably low, particularly when they were ingested as a part of a cereal-based meal (Narasinga Rao *et al.*, 1972). Thus, these insoluble iron phosphates, though satisfactory from the standpoint of stability, were unacceptable from the standpoint of iron availability.

Another insoluble iron compound that is white in color is ferrous citrate (monoferrous acid citrate), from which iron was shown to be as available as the iron from ferrous sulfate (Narasinga Rao *et al.*, 1978a). Although insoluble and considered stable to oxidation, ferrous citrate developed a greenish yellow color when added to crude common salt. However, ferrous citrate was found to be stable when mixed with dry materials like sugar or wheat flour, and it may serve as an iron source for fortification of such food articles. Reduced iron powder (Mallinckrodt), which is also insoluble, is considered to be a good source for iron fortification (Shah and Belonje, 1973). It was not stable when mixed with crude common salt; it developed a brown color readily. Observations on insoluble salts are summarized in Table II.

Since it was found that many iron compounds when used singly for fortification were unsuitable, either because they were unstable or because the iron they contained was unavailable, two alternative approaches were investigated. In one, an attempt was made to identify a chemical agent that would prevent discoloration when ferrous sulfate, ferrous ammonium sulfate, or ferrous citrate was used as an iron source. In another approach an attempt was made to identify a chem-

Table II
INSOLUBLE IRON SOURCES TESTED FOR SALT FORTIFICATION

Iron source	Stability of fortified salt	Bioavailability of iron in fortified salt
Ferric phosphate Ferric pyrophosphate Sodium iron pyrophosphate	Stable	Unacceptably low in presence of a meal
Ferrous citrate	Unstable—yellow color develops within 2 days	As good as ferrous sulfate
Reduced iron	Unstable—brown color develops within a day	Reported to be good depending on particle size

ical agent that would enhance iron absorption from a stable insoluble iron compound like ferric phosphate.

C. Use of Stabilizers

Several chemical compounds that may be expected to prevent color formation by complexing with iron were studied (Narasinga Rao and Vijaya Sarathy, 1975). A list of the compounds investigated is given in Table III. Of all the compounds tested, only orthophosphoric acid (OPA) and sodium hexametaphosphate (SHMP) were found to prevent color development. Similar results were obtained when ferrous ammonium sulfate or ferrous citrate was used as an iron source. The effect of mixing different proportions of SHMP or OPA to salt fortified with $FeSO_4$ was then investigated. Only when these coordinating agents were present in excess (2 mol/mol $FeSO_4$), that is, 3500 ppm of SHMP or OPA for 3500 ppm $FeSO_4$, did the salt remain stable without any color development when stored for long periods (8 months or more). For SHMP to act as an effective coordinating agent and prevent color, an acid medium was essential; addition of 1500 ppm of sodium acid sulfate ($NaHSO_4$) was found to be optimal for this purpose. Orthophosphoric acid could substitute for both SHMP and $NaHSO_4$. Other coordinating agents like sodium acid pyrophosphate or tetrasodium pyrophosphate could not replace SHMP. Athough salt fortified with $FeSO_4$ and one of the stabilizing agents just listed was without color, $FeSO_4$ underwent other chemical changes with time. In the salt fortified with SHMP and $FeSO_4$, iron was completely oxidized at the end of 18 weeks and had become almost insoluble. In salt fortified with OPA also, the levels of soluble iron and ferrous iron decreased to a low value during storage. Accompanying a decrease in soluble iron, bioavailability had also decreased. Presumably, ferrous iron is practically converted to ferric iron, which reacts with phosphate yielding insolu-

Table III

CHEMICAL ADDITIVES TESTED TO PREVENT COLOR FORMATION IN SALT FORTIFIED WITH FERROUS SULFATE[a]

Sodium acid sulfite	Glutamic acid
Sodium sulfite	Lysine
Sodium thiosulfate	Cystine
Sodium metabisulfite	Succinic acid
Ascorbic acid	Histidine
Sodium ascorbate	Sodium hexametaphosphate (SHMP)
Tocopherol	Sodium dihydrogen phosphate
Cysteine HCl	Sodium acid sulfate
Glucose	Orthophosphoric acid[b] (OPA) (3125 ppm)
Sodium EDTA	Sodium hexametaphosphate[b] (SHMP) (3125 ppm)
Citric acid	Sodium acid sulfate (1875 ppm)
Trisodium citrate	Cysteine–H_3PO_4[b]
Nicotinic acid	Citric acid–H_3PO_4[b]
Glycine	Ascorbic acid–H_3PO_4[b]

[a]Ferrous ammonium sulfate and ferrous citrate were also tested in place of ferrous sulfate, with similar results. The additives were tested in equimolar proportions. Only orthophosphoric acid and sodium hexametaphosphate could prevent color development.

[b]Fortified salt containing these additives remained colorless for 8 months or more. Iron bioavailability, though satisfactory, initially deteriorated on storage.

ble ferric phosphate. Iron availability in salt fortified with SHMP was low even initially (Narasinga Rao and Vijaya Sarathy, 1975). In the case of salt fortified with OPA, although satisfactory initially, availability deteriorated on storage (Narasinga Rao and Vijaya Sarathy, 1975).

D. Use of Absorption Promoters

In the second approach, an attempt was made to use an absorption promoter with a stable but poorly absorbed iron compound like ferric phosphate (Narasinga Rao and Vijaya Sarathy, 1975). A number of compounds so tested are listed in Table IV. Ascorbic acid actually developed a pink color when added to crude salt. A similar discoloration with cooking salt and ascorbic acid has been reported by Sayers and co-workers (1974). However, of the other three compounds—that is, cysteine hydrochloride, trisodium citrate, and sodium acid sulfate ($NaHSO_4$)—only with $NaHSO_4$ did the iron absorption from $FePO_4$ increase significantly; it nearly doubled, reaching a value 80% of that obtained with $FeSO_4$. Salt fortified with $FePO_4$ and $NaHSO_4$ in 1:2 molar proportions was found to be stable up to 8 months or more, and iron absorption from it did not deteriorate on storage. This formula was therefore proposed for salt fortification. According to the recommended formula, common salt is to be fortified with 3500

Table IV
CHEMICAL ADDITIVES TESTED TO IMPROVE IRON ABSORPTION FROM SALT FORTIFIED WITH FERRIC PHOSPHATE (1000 ppm Fe)

Chemical additive	Amount (ppm)	Stability	Increase in iron absorption from ferric phosphate
Ascorbic acid	3150	Pink color develops immediately	—
Cysteine HCl	3200	Stable for 3 months	Nil
Trisodium citrate	5300	Stable for 5 months	Nil
Sodium acid sulfate	5000	Stable for 8 months or more	Nearly twice

ppm $FePO_4$ (i.e., 1000 ppm iron) and 5000 ppm of $NaHSO_4$ to provide 1 mg iron/g fortified salt. Since the average consumption of salt in India is about 15 g/day per adult, use of fortified salt will provide 15 mg of additional iron daily.

Occasionally, with some batches of crude salt this formula gave a yellow coloration, but this color gradually disappeared within 7–10 days on exposure to the atmosphere. Presumably this happens with salt samples having a high content of $MgCl_2$. This can perhaps be avoided by a preliminary washing of salt to reduce the magnesium content, but the additional processing step would add to the cost of salt.

The recommended formula just given has been improved on (1) to avoid occasional yellow discoloration altogether and (2) to reduce the cost. Instead of using $FePO_4$, which is more expensive than $FeSO_4$, a mixture (equimolar proportions) of $FeSO_4$ (3200 ppm) and orthophosphoric acid (2200 ppm) along with $NaHSO_4$ (5000 ppm) can be used, and salt fortified according to this formula was found to be equally good (Narasinga Rao and Vijaya Sarathy, 1978b). One can use sodium orthophosphate (SOP) instead of OPA, since OPA, being a corrosive liquid, was found to present handling problems in large-scale fortification trials (R. N. Dutta, personal communication).

III. TECHNOLOGY OF SALT FORTIFICATION

A. Mixing Procedure

Fortification was first carried out on a laboratory scale. Since commercial salt is crystalline, it had to be ground to a coarse powder before mixing with the fortificants in order to ensure uniform mixing and distribution of iron. To 15 kg of crushed salt, 75 g of powdered sodium acid sulfate was added and mixed in a planetary mixer for 2 min. To this, 55 g of ferric orthophosphate was added and mixed for another 2 min. Uniformity of mixing was checked by estimating the iron content and pH of fortified salt samples. The iron ranged from 982 to 1110 ppm, and the pH of a 10% solution of the fortified salt ranged from 4.65 to 5.00.

The technology for the large-scale production of fortified salt has been standardized employing a ribbon blender (R. N. Dutta, personal communication). The powdered salt (40–50 mesh) is first mixed with OPA and then mixed with $FeSO_4$. The total mixing time is 7 min at a blender speed of 20 rpm. A screw-type conveyer mixer can also be used here with an appropriate feeder. Large-scale fortification can be done either by batch mixing or by continuous mixing. Large-scale fortification trials employing the recommended formula of $FePO_4$–$NaHSO_4$ or $FeSO_4$–OPA/SOP–$NaHSO_4$ have been carried out employing a batch mixer. The uniformity of mixing was found to depend on the size of the mixing unit, the quality of raw salt and the time of blending. When $FeSO_4$–OPA–$NaHSO_4$ is the formula used, the fortificants can be dissolved in a minimal quantity of water and sprayed over the raw ground salt.

B. Cost of Fortification

Because of escalating costs for machinery and chemicals, it may be hazardous to give specific figures. However, on the basis of prevailing prices in India (1982), the cost of machinery and incidental expenditure has been estimated to be around Rs7.0 lakhs* (U.S. $70,000) for a unit producing 12,000 tons of fortified salt per annum. The cost of processing fortified salt, excluding cost of chemicals, will be approximately Rs35 (U.S. $3.50) per ton. Cost of production, including the cost of chemicals and packing in special bags at the current prices, will be Rs250 (U.S. $25.00) per ton using $FePO_4$, and Rs160 (U.S. $16.00) per ton using $FeSO_4$–OPA. These costs may decrease when large-scale manufacture of fortified salt is undertaken and required chemicals are manufactured in bulk.

C. Packaging and Transportation

The ordinary gunny bags normally used for salt transportation in India cannot be be used for the fortified salt because of the corrosive nature of the chemicals used as well as the low pH of the fortified salt. Packing of fortified salt in unlaminated high-density polyethylene gunny bags was found satisfactory for storage and transportation over long distances.

*The Indian unit of currency is the rupee (Re, plural Rs); lakh is equivalent to 100,000.

IV. CONSUMER ACCEPTABILITY

Consumer acceptability trials have been carried out with salt fortified with $FePO_4$–$NaHSO_4$. One such trial was carried out in Hyderabad as soon as the formula was developed, by distributing the fortified salt to a number of families with different socioeconomic backgrounds. This fortified salt was found to be generally acceptable. When used in cooking it did not alter the taste or color of any of the preparations except in two or three dishes that normally change color when exposed to iron. A large acceptability trial in Delhi carried out by the Food and Nutrition Board of the government of India has also confirmed the acceptability of iron-fortified salt. Both in the community trial in schools and in field trials in four regions of the country, fortified salt was found to be generally acceptable and was compatible with cooking practices prevailing in different parts of the country.

V. IMPACT OF FORTIFIED SALT ON IMPROVING IRON STATUS OF THE COMMUNITY

A community trial was carried out among children aged 5–15 years, the inmates of residential certified schools, to test the efficacy of salt fortified with $FePO_4$–$NaHSO_4$ in correcting or preventing iron deficiency anemia (Nadiger *et al.*, 1980). Iron-fortified salt was supplied to the experimental school and unfortified salt to the control school for a period of 18 months. These schools were not buying salt from any other source during the study. The results of this study indicated that at the end of 1 year the mean hemoglobin level in children receiving the fortified salt had significantly increased, while in the control group receiving the unfortified salt, hemoglobin levels remained essentially unchanged. The prevalence of anemia also decreased in the experimental group but not in the control group (Nadiger *et al.*, 1980).

A more comprehensive field trial among rural populations was initiated in 1979 in four areas of the country, each covering 4000–6000 people and has since been completed. Fortified salt for these field trials was manufactured centrally and distributed. The results of the multicentric study (Working Group Report, 1982) indicate that fortified salt had a significant impact by increasing hemoglobin levels and correcting anemia in these rural populations also. The impact was the highest in a region where incidence of anemia was highest due to hookworm infestation. Deworming concomitantly had only a small additional benefit. The results of these community trials indicate that iron-fortified salt can be used as a public health measure to control and prevent anemia in India. Steps are currently being taken to implement the program of fortification of salt with iron to control anemia on a national scale.

VI. OTHER SALT FORTIFICATION TRIALS

Attempts are also being made to use iron-fortified salt in Indonesia and Thailand to control iron deficiency anemia. In Thailand a formula based on SHMP and $NaHSO_4$ that was tested earlier at Hyderabad is being tried (Suwanik *et al.*, 1980). The crude salt is subjected to a purification step before fortifying. The iron absorption from a meal containing this fortified salt is reported satisfactory. Field trials with this salt are reported to be under way in Thailand, but results of this study are not yet available. Bioavailability of iron from salt fortified with $FeSO_4$–SHMP–$NaHSO_4$ was reported earlier by us to decrease on storage because of formation of insoluble ferric phosphate. The efficacy of salt fortified according to this formula therefore needs to be carefully evaluated on the basis of results of field trials before it is recommended for large-scale fortification. In some countries like Sri Lanka, cooking salt is steeped in water, and only the saturated brine is used for cooking. In such cases, fortification of salt with iron may not be feasible.

VII. USE OF IRON AND IODINE IN SALT FORTIFICATION

Salt has also been used in many countries as a vehicle for iodine fortification. The question of fortifying salt with both iron and iodine is being considered by several countries. Attempts are being made at Hyderabad to incorporate iodine into salt fortified with iron ($FePO_4$–$NaHSO_4$). Some preliminary studies indicated that neither potassium iodide nor potassium iodate may be very stable in the iron-fortified salt. Systematic studies to fortify salt with iron and iodine simultaneously are currently in progress at Hyderabad. Attempts are also being made in Thailand and Indonesia to fortify salt with both iodine and iron.

REFERENCES

International Nutritional Anemia Consultative Group. (1977). "Guidelines for the Eradication of Iron Deficiency Anaemia." Nutrition Foundation, New York.

Nadiger, H. A., Krishnamachari, K. A. V. R., Nadamuni Naidu, A., Narasinga Rao, B. S., and Srikantia, S. G. (1980). The use of common salt (sodium chloride) fortified with iron to control anaemia: Results of preliminary study. *Br. J. Nutr.* **43,** 45–51.

Narasinga Rao, B. S., Surendra Prasad, and Apte, S. V. (1972). Iron absorption in Indians studied by whole body counting: A comparison of iron compounds used in salt fortification. *Br. J. Haematol.* **22,** 281–286.

Narasinga Rao, B. S., and Vijya Sarathy, C. (1975). Fortification of common salt with iron: Effect of chemical additives on stability and bioavailability. *Am. J. Clin. Nutr.* **28,** 1395–1401.

Narasinga Rao, B. S., Soonita Kathoke, and Apte, S. V. (1978a). Mono ferrous acid citrate ($FeC_6O_7 \cdot H_2O$) as an iron fortificant. *Br. J. Nutr.* **39,** 663–665.

Narasinga Rao, B. S., and Vijaya Sarathy, C. (1978b). An alternate formula for the fortification of common salt with iron. *Am. J. Clin. Nutr.* **31,** 1112–1114.

Sayers, M. H., Lynch, S. R., Charlton, R. W., and Bothwell, T. H. (1974). Iron absorption from rice meals cooked with fortified salt containing ferrous sulphate and ascorbic acid. *Br. J. Nutr.* **31,** 367–375.

Shah, B. G., and Belonje, B. (1973). Bioavailability of reduced iron. *Nutr. Rep. Int.* **7,** 151–156.

Suwanik, R., and The Study Group, Bangkok. (1980). "Iron and Iodine Fortification in Thailand." Faculty of Medicine, Siriraj Hospital, Mahidol University, Bangkok.

Working Group Report. (1982). Use of common salt fortified with iron in the control and prevention of anaemia—A collaborative study. *Am. J. Clin. Nutr.* **35,** 1442–1451.

13
Condiments

LARS GARBY
University of Odense
Institute of Physiology
Odense, Denmark

I. INTRODUCTION

Two different condiments, fish sauce and monosodium glutamate (MSG), both widely used in East and Southeast Asia, have been suggested and partly evaluated as vehicles for iron fortification.

II. FISH SAUCE

A. Production and General Characteristics

Fish sauce is a clear light-brown to brown liquid food product prepared by fermentation and salt extraction of marine fish. The duration of extraction, taking

ISBN 0-12-177060-5

place at ambient temperature, is generally several months, although attempts have been made to enhance the rate of extraction by addition of enzymes. The mixture (macerate) is filtered, and the filtrate is diluted with salt water to obtain different qualities.

The production technique and several physical and chemical characteristics of fish sauce have been studied and described (Food and Agriculture Organization, 1967). The salt content is high, about 200–300 g/liter. Depending on the quality—that is, the degree of dilution—fish sauce contains small amounts of peptides and amino acids. It is slightly acidic, pH about 5–6, with a very low buffer capacity.

Several very cheap products are now being marketed under the name of fish sauce. They contain minute amounts of fish extract and are essentially simple salt solutions to which coloring and flavoring materials have been added (burnt sugarcane, extracts from buffalo skin, etc.). Also, to improve the taste, MSG is added.

B. Use, Consumption, and Costs

Fish sauce has been used in Southeast Asia for as long as anyone can remember. It is widely used as a condiment, a flavoring material, and, perhaps most importantly, a substitute for solid salt. It is added during cooking and to the meals to most kinds of food.

A very large percentage of population groups in Kampuchea, Thailand, and Vietnam cover their daily requirements of salt in the form of fish sauce. This also applies to a lesser extent to population groups in Malaysia, the Philippines, Indonesia, and Burma.

The daily consumption of fish sauce in populations that consume it varies between 10 and 40 ml. The cost of fish sauce in towns and village shops in Thailand depends on the quality and varies between U.S. $0.20 and $1.00/liter; the undiluted high-quality sauce probably makes up only a very small fraction of the total quantity sold.

In several regions of Southeast Asia (e.g., Thailand), by far the most fish sauce consumed is produced in large factories. In other regions, local production at the household level predominates.

C. Effects of Addition of Iron Salts on Visual Appearance of Fish Sauce

Garby and Areekul (1974) have reported on effects on the visual appearance of fish sauce with addition of several different iron salts. The following iron (Fe) compounds were added to a final concentration of 0.5 and 1.0 mg of elemental iron per milliliter of fish sauce: Fe(III) sodium ethylenediaminetetraacetate (FeNa EDTA), Fe(III) choline hydroxide citrate, Fe(II) glycine sulfate, Fe(II) digluconate, Fe(II) succinate, and Fe(II) sulfate. The appearance of the fish sauce was observed at room temperature (22–24°C) for several weeks in stoppered test tubes. No change in the visual appearance took place in the samples containing FeNa EDTA, whereas precipitation took place in the other samples within a few minutes to hours.

Precipitation after addition of FeNa EDTA has been reported in some marketed brands of fish sauce. This phenomenon may be associated with artificial coloring and flavoring substances in these brands, but the problem needs further study.

D. Effect of Addition of Fe(III) Sodium EDTA on Taste of Various Foods with Addition of Enriched Fish Sauce

The effect of addition of FeNa EDTA to fish sauce on the taste of various foods to which the enriched fish sauce was added was studied by Garby and Areekul (1974). The meals were typical of Thailand and presumably of most of Southeast Asia. A double-blind technique was used, and no effect was found.

E. Absorption of Iron from Foods Flavored with Fish Sauce Enriched by FeNa EDTA

Garby and Areekul (1974) measured the absorption of iron from local (Thai) meals to which enriched fish sauce (5 mg Fe/meal as FeNa EDTA) was added in a number of subjects recovering from mild disorders (parasitic infections). The average absorption percentage was 7–8%. This figure is in good agreement with later, much more extensive data on the absorption of iron (added as FeNa EDTA) from various meals (Martinez-Torres *et al.*, 1979; L. Hallberg, personal communication), showing, in general, that the iron in FeNa EDTA is better absorbed than ferrous sulfate.

F. Effect of Enriched Fish Sauce on Hematocrit Values in a 1-Year Pilot Field Fortification Trial

A pilot field fortification trial on the effect of adding 1 mg iron (as FeNa EDTA) per milliliter of fish sauce for 1 year was carried out in a province in the central part of Thailand (Garby and Areekul, 1974). The response was measured in terms of increases in hematocrit values in comparison with a control population. A response was found in 25–35% of the subjects, and the mean response in these subjects was about 4.5 hematocrit units, or about 15%. This statistically significant response must be judged to be clinically significant as well.

A large-scale field fortification trial was started in Thailand in 1980.

G. Methods and Costs of Fortification of Fish Sauce

In countries like Thailand, where by far most of the fish sauce is produced in 50–100 large factories, the method of fortification presents no problems. The very stable FeNa EDTA salt is simply added directly to the sauce before bottling. In regions where there are many small production units, the logistical problems may be considerable.

Technical-grade FeNa EDTA is presumably sufficiently pure for fortification of fish sauce. The price is U.S. $5–10/kg when bought in large quantities, that is, about U.S. $1/kg of elemental iron or U.S. $0.01 per "fortification unit" (10 mg Fe). The chemically pure substance is about twice as expensive.

III. MONOSODIUM GLUTAMATE

A. Production and General Characteristics

Monosodium glutamate, the sodium salt of the naturally occurring glutamic acid, is produced microbiologically and is obtained as the monohydrate in crystalline needles or powder form. The salt is white and highly soluble in water.

The world production in 1976 was about 230 kilotons, most of it for human use.

B. Use, Consumption, and Costs

Monosodium glutamate is used practically throughout the world as flavoring material and is added during cooking and to most kinds of food.

In many population groups in East and Southeast Asia, MSG consumption is considerable. In the Philippines, some 95% of the population consumes MSG daily. It provides to some extent the physiological requirement for sodium.

The wholesale price of MSG is U.S. $1–2/kg. In several countries in East and Southeast Asia it is marketed in labeled cellophane packets containing about 2 g of the salt, that is, roughly the daily consumption of a family.

C. Effect of Addition of Iron Compounds on Some Physical and Chemical Properties of Monosodium Glutamate

The source of iron to be added to MSG should be white or nearly white in color. Bauernfeind and Timreck (1978) have reported results of addition of two iron compounds: ferric orthophosphate and zinc stearate-coated ferrous sulfate.

Ferric orthophosphate was obtained from Joseph Turner Co. (Ridgefield, New Jersey) and described as a 99% 325-mesh, off-white powder (phase II), odorless, tasteless, and meeting all U.S. state and federal requirements. The particle size was consistently <1 μm for 90% or more of the material and, on this basis, expected to have an absorbability about 50% of that of ferrous sulfate.

Zinc stearate-coated ferrous sulfate was obtained from Durkee Industrial Foods Group (SCM Corp., Cleveland, Ohio). It is described as a light-colored, tasteless, relatively nonreactive, free-flowing powder varying in particle size, the majority of which is 100–200 mesh. The coating protects the ferrous sulfate up to a temperature of 122°C, the melting point of the encapsulating material. The biological value (i.e., its relative absorbability) in rats in comparison with ferrous sulfate is about 70%.

When the equivalent of 50 mg of elemental iron of each of the compounds were mixed into packets containing 2.4 g MSG, it was found that reasonably uniform packaged iron content was achieved with either iron source. Furthermore, these mixtures did not appear to influence significantly the distribution of vitamin A that was also added to the packets.

Further studies on the effect of the mentioned iron compounds on the stability of added vitamin A showed no adverse effects.

D. Acceptability of the Fortified Vehicle and Absorption of Added Iron

Studies of the acceptability of the fortified vehicle are in progress in the Philippines. Studies of the absorption of iron from meals to which the fortified vehicle has been added have not yet been carried out.

E. Methods and Costs of Fortification of Monosodium Glutamate

There appear to be no technical problems with respect to the fortification of MSG with the iron compounds mentioned before. The costs of the iron sources are at present about U.S. $0.75/kg for ferric orthophosphate and U.S. $2.27/kg for the equivalent coated ferrous sulfate.

REFERENCES

Bauernfeind, J. C., and Timreck, A. (1978). Monosodium glutamate, a food carrier for added vitamin A and iron. *Philipp. J. Sci.* **107**(3–4), 203.

Food and Agriculture Organization. (1967). "Fish Processing in the Indo-Pacific Area," Indo-Pac. Fish. Counc. Reg. Stud. No. 4. FAO Regional Office for Asia and the Far East, Bangkok, Thailand.

Garby, L., and Areekul, S. (1974). Iron supplementation in Thai fish sauce. *Ann. Trop. Parasitol.* **68,** 467.

Martinez-Torres, C., Romana, E. L., Renzi, M., and Layrisse, M. (1979). Fe(III)–EDTA complex as iron fortification, Further studies. *Am. J. Clin. Nutr.* **32,** 809.

Index

FOOD SCIENCE AND TECHNOLOGY
A SERIES OF MONOGRAPHS

Maynard A. Amerine, Rose Marie Pangborn, and Edward B. Roessler, PRINCIPLES OF SENSORY EVALUATION OF FOOD. 1965.

Martin Glicksman, GUM TECHNOLOGY IN THE FOOD INDUSTRY. 1970.

L. A. Goldblatt, AFLATOXIN. 1970.

Maynard A. Joslyn, METHODS IN FOOD ANALYSIS, second edition. 1970.

A. C. Hulme (ed.), THE BIOCHEMISTRY OF FRUITS AND THEIR PRODUCTS. Volume 1—1970. Volume 2—1971.

G. Ohloff and A. F. Thomas, GUSTATION AND OLFACTION. 1971.

C. R. Stumbo, THERMOBACTERIOLOGY IN FOOD PROCESSING, second edition. 1973.

Irvin E. Liener (ed.), TOXIC CONSTITUENTS OF ANIMAL FOODSTUFFS. 1974.

Aaron M. Altschul (ed.), NEW PROTEIN FOODS: Volume 1, TECHNOLOGY, PART A—1974. Volume 2, TECHNOLOGY, PART B—1976. Volume 3, ANIMAL PROTEIN SUPPLIES, PART A—1978. Volume 4, ANIMAL PROTEIN SUPPLIES, PART B—1981. Volume 5, SEED STORAGE PROTEINS—1985.

S. A. Goldblith, L. Rey, and W. W. Rothmayr, FREEZE DRYING AND ADVANCED FOOD TECHNOLOGY. 1975.

R. B. Duckworth (ed.), WATER RELATIONS OF FOOD. 1975.

Gerald Reed (ed.), ENZYMES IN FOOD PROCESSING, second edition. 1975.

A. G. Ward and A. Courts (eds.), THE SCIENCE AND TECHNOLOGY OF GELATIN. 1976.

John A. Troller and J. H. B. Christian, WATER ACTIVITY AND FOOD. 1978.

A. E. Bender, FOOD PROCESSING AND NUTRITION. 1978.

D. R. Osborne and P. Voogt, THE ANALYSIS OF NUTRIENTS IN FOODS. 1978.

Marcel Loncin and R. L. Merson, FOOD ENGINEERING: PRINCIPLES AND SELECTED APPLICATIONS. 1979.

Hans Riemann and Frank L. Bryan (eds.), FOOD-BORNE INFECTIONS AND INTOXICATIONS, second edition. 1979.

N. A. Michael Eskin, PLANT PIGMENTS, FLAVORS AND TEXTURES: THE CHEMISTRY AND BIOCHEMISTRY OF SELECTED COMPOUNDS. 1979.

J. G. Vaughan (ed.), FOOD MICROSCOPY. 1979.

J. R. A. Pollock (ed.), BREWING SCIENCE, Volume 1—1979. Volume 2—1980.

Irvin E. Liener (ed.), TOXIC CONSTITUENTS OF PLANT FOODSTUFFS, second edition. 1980.

J. Christopher Bauernfeind (ed.), CAROTENOIDS AS COLORANTS AND VITAMIN A PRECURSORS: TECHNOLOGICAL AND NUTRITIONAL APPLICATIONS. 1981.

Pericles Markakis (ed.), ANTHOCYANINS AS FOOD COLORS. 1982.

Vernal S. Packard, HUMAN MILK AND INFANT FORMULA. 1982.

George F. Stewart and Maynard A. Amerine, INTRODUCTION TO FOOD SCIENCE AND TECHNOLOGY, second edition. 1982.

Malcolm C. Bourne, FOOD TEXTURE AND VISCOSITY: CONCEPT AND MEASUREMENT. 1982.

(continued)

R. Macrae (ed.), HPLC in Food Analysis. 1982.

Héctor A. Iglesias and Jorge Chirife, Handbook of Food Isotherms: Water Sorption Parameters for Food and Food Components. 1982.

John A. Troller, Sanitation in Food Processing. 1983.

Colin Dennis (ed.), Post-Harvest Pathology of Fruits and Vegetables. 1983.

P. J. Barnes (ed.), Lipids in Cereal Technology. 1983.

George Charalambous (ed.), Analysis of Foods and Beverages: Modern Techniques. 1984.

David Pimentel and Carl W. Hall, Food and Energy Resources. 1984.

Joe M. Regenstein and Carrie E. Regenstein, Food Protein Chemistry: An Introduction for Food Scientists. 1984.

R. Paul Singh and Dennis R. Heldman, Introduction to Food Engineering. 1984.

Maximo C. Gacula, Jr., and Jagbir Singh, Statistical Methods in Food and Consumer Research. 1984.

S. M. Herschdoerfer (ed.), Quality Control in the Food Industry, second edition. Volume 1—1984. Volume 2 (first edition)—1968. Volume 3 (first edition)—1972.

Y. Pomeranz, Functional Properties of Food Components. 1985.

Herbert Stone and Joel L. Sidel, Sensory Evaluation Practices. 1985.

Fergus M. Clydesdale and Kathryn L. Wiemer (eds.), Iron Fortification of Foods. 1985.

In preparation

Robert V. Decareau, Microwaves in the Food Processing Industry. 1985.

S. M. Herschdoerfer (ed.), Quality Control in the Food Industry, second edition. Volume 2—1985. Volume 3—1986. Volume 4—1987.